The Truth About Mars

An Eyewitness Account

Other Books by Ernest L. Norman

The Infinite Concept of Cosmic Creation

Tempus Interludium, volume one

Tempus Interludium, volume two

The Voice of Venus

Infinite Contact

Infinite Perspectus

Cosmic Continuum

Tempus Invictus

Tempus Procedium

Anthenium

Elysium

The Truth About Mars

An Eyewitness Account

ERNEST L. NORMAN

Fourth Edition

UNARIUS ACADEMY OF SCIENCE
El Cajon, California

The Truth About Mars

An Eyewitness Account

 For information address: The Unarius Educational Foundation, 145 S. Magnolia Avenue, El Cajon, CA 92020. E-mail: uriel@unarius.org www.unarius.org.

Library of Congress Catalog-in-Publication Data

Norman, Ernest, L.
The Truth About Mars: An Eyewitness Account

Includes bibliography
ISBN 0-935097-39-2
Library of Congress Catalog Number 97-61481

1. Life on Mars 2. Martian Civilization
3. Science & Technology

Title I

Printed in the United States of America

Ernest L. Norman

Pioneering Researcher

In the New Science of Interdimensional Physics

Founder, Unarius Academy of Science

Table of Contents

Foreword

During recent months the Red Planet, Mars, the fourth planet from our sun, has generated more interest among earth scientists, philosophers, science fiction writers, artists, musicians, and the public as a whole than any other planet in our solar system has generated during our written history, from early Egyptian times to the present year of 1998.

Authors have out-pictured the thoughts and feelings that live in the consciousness of those people who are witness to their memories of the past, when we of Earth and the people of Mars communicated via spacecraft and via more advanced methods of telecommunication which are based upon the interdimensional physics of consciousness.

No, *The Truth About Mars* is not science fiction, nor is it a book about the possibility of life on an extraterrestrial planet. And why is this not fiction, you may ask?

Forty-three years ago, in 1955, Ernest L. Norman visited the underground cities of Mars, and on thirty-three separate occasions he recorded his findings. His eyewitness accounts describe the Martian people and an advanced social system that is built upon scientific principles which are far in advance of those understood on Earth. These visits also provided him with evidence to explain the present geological condition of the planet and the reason that the surface of Mars is destitute.

Martian scientists gave him the understanding that their planet had at one time been similar to Earth and, like Earth, it had once abounded in oceans, rivers, and lush vegetation. In fact, in its geological composition, Mars is a sister planet to Earth.

The Truth About Mars answers the question: Is there human life on other terrestrial planets in our galaxy? It also addresses the nature of consciousness and man's ability to make use of the electronic design of the mind in the exploration of the universe.

Ernest L. Norman, a scientific researcher in the paranormal, and most importantly a researcher in the physics of consciousness, demonstrates that the mind is, in actuality, a fourth-dimensional energy device that functions in a manner that is similar to a television broadcasting system. It serves to transmit information on high-frequency carrier waves, which are then demodulated to a receiver who activates the signal on a lower frequency, so it can then be viewed as sound and image.

In this way Dr. Norman learned that the Martian

people are an advanced race of humankind. Thousands of years ago they had developed the science of astrophysics so that they were able to see into the cosmos and recognize the evidence of a supernova which would have dire consequences for the people living on their planet. Because of their knowledge and ability to see into the future, they were able to build underground cities to protect the population. The consequences of the explosion of the supernova removed 90% of the oxygen from the atmosphere of Mars, and it had other consequences that made the surface of Mars a desert.

There is now evidence of ancient substructures on Mars. With the advanced photographic systems of *Surveyor,* we should see evidence that will reveal the Face on Mars, and other anomalies will also be resolved, attesting to what is evidence of an ancient civilization.

This book however is not about the paranormal nature of life; it is the statement of one man's experience, validated by thousands of photographs returned by the *Mariner, Viking, Pathfinder,* and *Global Surveyor* space probes.

Ten appendices support the statements of the author. They verify that this book *The Truth About Mars* is not the product of one man's imaginings, but that the author's mind functioned like a zoom lens, recording information that originated thousands of miles from its destination. His mind was able to function as did the camera that recorded the photographic images which have come from the space probes. Ernest Norman, then, can be considered to be a role model, demonstrating what can be achieved when the mind is understood to be an electronic device capable of functioning interstellarly and interdimensionally.

And what can be said about the present mental state of the people on planet Earth who find it difficult to conceive that life as we know it on our terrestrial planet is also experienced by people similar to us, and who, while occupying a five-sense physical anatomy, are also concerned with the meaning of life's equation?

The news about present research being conducted by NASA is electrifying, as are the accelerated programs to investigate the possibility of life on Mars. These programs are making headlines in the media because there is strong public interest in our neighboring planet. This interest of course has been developed because all individuals contain, in their subconscious or unconscious memories, evidence of previous experiences, memories of previous interactions between the people of Mars and the people of Earth.

The avalanche of information pressing into the minds of people who have been questing about the evolutionary design of life, during the forty years since the first Martian probe, *Mariner,* has increased a thousand-fold because of advanced computer technology and aerospace engineering. This is evidenced in the *Mars Global Surveyor*'s cameras, which are taking enhanced photographic pictures of the Martian terrain. They reveal that water had at one time been extensive on the surface of the Martian planet.

Clearly, the purpose of this book is to advance our knowledge not only of the evidence of another terrestrial civilization that has the greatest similarity to Earth, but it is also to advance the greater understanding of the nature of consciousness. The propulsion systems, which are used in the technology applied for the exploration of space and the discovery of intelligent life in the universe, and the nature and function of the mind are not far apart. In fact, the mind is the propulsion system that serves to expand one's consciousness in order to explore the frontiers of our present environment and the planet on which we live to learn about the infinite nature of life.

We as human beings have been searching for the evidence of extraterrestrial intelligence, as evidenced by the SETI and Phoenix programs. We are not alone in the universe. The existence of other humanity on nearby planets will be revealed through space exploration, but, most importantly, humankind is being opened to the new frontier of extraterrestrial civilizations.

Yet the basic inquiry requires man's recognition that he carries within himself the substance which is recorded in so-called space and time. When we can assimilate this, it will serve to remove the amnesia that has blocked our memory of the reality of extraterrestrial life on other than one earth planet, in one solar system, in one small galaxy, in one universe.

Humankind will resolve the enigma of life and death and his purpose in the cosmic design that we see in the hundreds of billions of star systems within our galaxy. *The Truth About Mars* is a forerunner for other books that have been written by Ernest L. Norman about the interdimensional physics of life. They serve as a radio-telescope outlining our past, present, and future.

Dr. Charles L. Spiegel
El Cajon, California
February 28, 1998

Preface

For centuries, before spacecraft transformed our understanding of the solar system, almost all the planets were believed, with varying degrees of conviction, to be inhabited. This belief conformed to the Copernican worldview, in which Earth is not unique among the planets. When the Space Age opened in 1957, the evidence for indigenous life was strongest for Mars—almost conclusive, on the face of it—and from the beginning the search for life on Mars became a major goal of the United States Space Program. The result was the historic series of spacecraft that began with *Mariner 4*, in 1965, and ended with *Vikings 1 and 2,* in 1976.

The dawning of the Space Age is a material event, occurring in the latter part of the 20th century, and as the description that follows is the result of a developed mental means of communication, it strengthens the understanding of this means of communication as the real and more apparent factor in the transference of energy from point to point.

Dr. Ernest L. Norman is the progenitor and a mental giant, writing from the higher statement of a full consciousness of the reality of the mind as a cosmic centrifuge. This is the first book, written by him in 1955, inaugurating the Unarius Academy of Science.

In this small but important book—a first—Dr. Norman extends the present knowledge of life on our close companion planet Mars, which has all of the basic biological and geological conditions for the evolutionary development of life forms, including Homo sapiens. He addresses mankind, not with the hardware of rocketry, space probes, and other sophisticated, high-energy technology, but he introduces the individual to the probability of life on other planets. Through the higher factors of the mind, he immediately takes the reader to planet Mars to reveal the enigma of this planet. This enigma has been tantalizing the minds of humankind for hundreds of years, resulting in a series of narrations and books, which over the years titillated the imagination of mankind until the present time.

The Truth About Mars describes how the power of the mind can be applied in all ways necessary for the individual to expand the central intelligence of himself in his relationship to the vitality of society.

For this reason, Dr. Norman points the way for aspiring mankind to overcome the limitations of his confinement in the material sectors of life. The pioneering efforts of our scientific community, through the auspices of the United States National Aeronautics & Space Administration, are serving as a preparatory basis for the recognition of extraterrestrial civilizations.

The expository nature of Dr. Norman's description of the underground cities of Mars is a vital teaching, not only of the magnanimous nature of man in overcoming the limitations of his material environment but also of the reality of interplanetary life in the universe.

The data now being collated and the many thousands of computer-enhanced images from the *Viking, Galileo, and Pathfinder* probes are validating the prehistory of Mars as an integrated and formed world of an ancient civilization. However, it will not be until earth humans land on Mars that the evidence of Martian society will be known and accepted.

The computer-enhanced image revealing a Face on Mars is capturing the interest of many people, including those in the scientific sector; this is a prelude to the buildup of further hard data, revealing the evidence of the ruins of ancient, advanced civilizations in the Cydonia region of Mars.

All of the precluding is the material evidence necessary for those who do not have second sight and who have not developed the reality of their sixth, seventh, and higher senses. But this should not deter anyone's search for the truth of extraterrestrial life. In fact, it is the objective of this writing to wake up the reader to his unlimited capacity in that he, too, can communicate through his sixth and higher senses. This so-called telepathy, a principle of frequency and harmonics, will reveal to him the evidence of the evolutionary design of life on companion earth planets.

The exploration of Mars is, in this sense, the first phase of reawakening us to the reality of the expanding universe, right in our own backyard, in our own solar system. The reality is that earth planets are abundant in our Milky Way Galaxy. The reality is that man has communicated via spacecraft and is continuing to communicate with the denizens of this planet through telecommunication or mental means—the normal methods of communication of extraterrestrial space brothers.

With all of this, of course, is the other reality! Present earth history has not revealed the true story of man's past and the reasons technology has outdistanced the knowledge of man's humanity to his fellow man and his creative spirituality.

You will therefore read in this book that the Martian people are wiser in their knowledge of their inherent human nature, and they practice this cosmic principle in their society. War is not a factor to disrupt their peaceful existence, and science and technology function equally balanced with man's spiritual coefficient.

The picture that emerges is of earthman beginning to realize that he has the capacity to leave his nest, to move out into the stars and regain his lost heritage, the birthright which has always been the objective and purpose for his being. Essentially, life's purpose is to advance one's knowledge of the principle of evolution, to integrate these evolutionary principles, using both the material technology and the spiritual technology, thus advancing both man and society on an upward and intelligent, progressive evolution.

Dr. Ernest L. Norman is the cofounder of the Unarius Educational Foundation. From 1954 to 1971, with his wife Ruth Norman, he pioneered the teaching of interdimensional physics, explaining the evolutionary nature of atoms and man, planets and solar systems, galaxies and universes. He authored the first twenty texts of the curriculum of the Unarius Academy of Science—teachings which explain the grand unified field theory and quantum mechanics. It is a cosmology that unites the mind and the universe to explain the nature of physical reality.

Introduction

A few months ago the astronomical and astrophysical world was tremendously excited by the approaching conjunction of Earth with the planet Mars. Many prominent astronomical authorities hoped to settle, once and for all time, the old, controversial issues about this planet: i.e., were there canals or were there not canals on Mars, and was this planet inhabited by some form of human life? So far as can be ascertained, the results of these investigations, after thousands of photographs of the planet and numerous controversies, were that some groups were even more firmly entrenched in their original ideas, while others became more confused than ever. In an overall sense, it can be said that they arrived at no definite conclusion whatsoever, nor will any new conclusions,

The author, writing in May, 1955, refers to the conjunction of Mars with Earth at the turn of the year, 1954 to 1955. See Addendum also.

which may be arrived at by the same token, have any more validity than the original concepts.

Therefore any new attempt at visualizing life or the canals on Mars resolves into the realm of clairvoyance and not by the making of bigger and more powerful telescopes. The two hundred-inch telescope at Mt. Palomar has, in a sense, merely increased the size of the universe for man rather than brought it closer to him. Now, just in case the term clairvoyant should arouse any antagonism or question in some persons, let us digress a moment to explain just what is meant by the word *clairvoyance.*

Clairvoyance, or the development of the sixth sense, is only another word for extrasensory perception taking place within the consciousness of man rather than through the reactionary physical senses such as physical sight, hearing, smell, touch, and taste. Practically everyone on this earth has had or will, at some time, be in a semi or momentarily clairvoyant state. Anyone having a so-called hunch or premonition is momentarily in such a state. At the great Duke University, Dr. Rhine, in his fifty years of research on parapsychology, has definitely established the facts and truths of extrasensory perception. There are numerous societies, associations, and organizations, national and international, whose many years of work have proven beyond a doubt that man does have, and can also develop, this extra sense or clairvoyance, sometimes called the sixth sense.

Any doubting Thomas can, if he is open minded enough, find, in a short time, an overwhelming mass of evidence to support this truth. In an advanced state of clairvoyance an individual sees and lives in a state of consciousness that "tunes" him in, as it were, to past or future events, places, and happenings, distance being no barrier. Such an awareness or consciousness is almost as real as the everyday objects around him. Any particular electrical or mechanical device, such as the television set, will demonstrate to some extent the nature of this conception.

A savage in the jungle would immediately be confounded were he presented with the appearance of some of our everyday appliances. He would quite likely, in his failure to understand or conceive, throw up his hands and deny the whole thing. He would, in a sense, be like the farmer who, having seen a giraffe for the first time, was exclaiming "there is no such animal!"

It is, as it has always been, the great lack of ability to form new concepts that has always caused man to throw up his hands and to cry out loudly against the appearance of any new thought, or mechanical or electrical contrivance. This is also quite true of most of our modern-day scientists, whether he is a man of medicine, of chemistry, or of astronomy. Therefore, it is up to individuals who have developed this extra sense or clairvoyance to fill in the obvious gaps in our many branches of science as well as in some of the more firmly-established spiritual concepts.

In writing an article of this kind, no effort is being made to prove what I have found and believe is true. Truth is entirely independent of the individual. After my thirty-five years of active research in the fields of electronics, physics, astrophysics, parapsychology, and their allied and associated sciences, I have succeeded in correlating and establishing an integrated concept which, when combined with a natural and highly-developed clairvoyance, gives, to a practically perfect degree, a direct mental contact not only person to person upon this earth but in contacting individuals living on other planets. This is, in a sense, what can be called conscious astral flight, inasmuch as I see the cities and the people and hear the individuals as they now exist on other planets. No mechanical devices are used, nor is any particularly advanced degree of trance state entered into. I maintain conscious continuity and am able to quote, at the time, just what is taking place, both audibly and visibly, at all times.

Contact With Planet Mars

1

Contact With Planet Mars

Since the dawn of time and the beginning of man's history on this planet, the starry skies have always been something of not only great and wondrous beauty, but the skies also have been filled with mystery and awe. The histories of ancient times contain numerous references to the sun, the moon, and the many bright stars and planets. Man has worshipped these heavenly bodies as deities or gods. Quite often the very nature of the religious beliefs of the peoples were woven around these mystical, heavenly orbs. Naturally, there has been a great deal of conjecture as to life and the existence of man in some form on some of these bright specks of light. This is especially true of the moon and of the planets of our solar system. Men like Copernicus, Galileo, Plato, etc., all speculated on this possibility.

During the last twenty-five years or so, there has been a tremendous impetus given to astronomical interests; perhaps this is partly due to the approach of the conclusion of a great cycle and the actual beginning of the Aquarian Age. There are numerous monthly publications that deal in a fictional way with interplanetary travel and life on other planets. There likewise are other articles and stories that have appeared from time to time dealing with flying saucers and spaceships, etc., which claim to be true, and as a small lad I shared this common interest in the heavens. Often I would peer through my father's telescope, which was of very modest power, at the moon or other bright points of interest; winter nights would often be devoted to pouring over any book or article containing anything of an astronomical nature.

It was not, however, until the close of World War II and the sudden influx of flying saucer stories that time and circumstances permitted resuming this fascinating subject. Along with metaphysical work, which I did both in churches and independently, the planets, space travel, etc., all became an integrated part of this work.

It was inevitable that sooner or later I should actually take a "flight" to some planet, not that this would be done in a rocket or some such machine; man has not progressed to such an advanced state of engineering as yet. So any such trips would be in a clairvoyant state. I am not the only one, by far, who has had such experiences; the persons both known and unknown who have made such flights and contacts are too numerous to mention at this time. I might add that much of what is written in the following pages has since been corroborated by some of these persons, without my previously having read any articles so written by them.

It has been my consistent habit to spend an hour or so of the late evening time in meditation. During these hours I have made innumerable contacts with those who have passed from this plane of existence. However, no serious attempt at interplanetary contact was tried until the second month of the year of 1955. At that time I began to be increasingly aware that something like this was being attempted by the peoples of other planets.

One evening, about the first part of May, of this year, while in a deep meditative state, I suddenly perceived a rather strange looking man standing before me. At first I thought him to be Chinese, as his dress and general appearance was somewhat similar to that of a man of ancient China. After introducing himself as Nur El, however, he quickly explained that he was from the planet Mars, and if I so desired, I could go there with him to his city, in astral flight, and that he would be my personal guide. He explained that his people were very desirous, in view of all the controversy going on, to clear up some of the so-called mysteries of Mars. He further assured me that it was quite obvious that a complete understanding was not possible in one visitation; therefore, as the first contact was made, it would be comparatively easy to establish other contacts, as was convenient and necessary.

Since this first contact and trip was made, I have

returned on several occasions; in fact, Nur El often stood beside me as I wrote to further clear up or refresh my memory regarding any details which were not entirely clear.

An Astral Flight Through Space

2

An Astral Flight Through Space

Now I will contact my Martian guide and take an astral flight through space, and see just how it is that man lives on the red planet. Almost immediately a very distinguished looking man stands before me; he is Nur El, a man of high position and esteem from one of the Martian cities. He is dressed in a very brilliant red suit. The coat is long, almost to the knees, with loose fitting pantaloons. On his head is a red hat with a square-shaped brim, which is turned up on four sides.

Our trip there is a matter of split seconds as no craft is used or needed. Arriving on the surface of Mars, we are at once aware of the extremely rugged terrain, rocky hills and sandy wastes that stretch out endlessly around us. There are many peculiar whirling dust clouds all about. Nur El explains that the ionosphere is very thin, which leaves the surface almost unprotected from the various beta, gamma, and cosmic rays. This high concentration of rays ionizes the very rare and gaseous atmosphere, and together with the thermal currents, creates terrific dust storms. There is also a very thinly divided dust layer on the ionosphere, which helps create the reddish appearance of the planet. There are a number of volcanoes, three of which are of major size; one of these was just barely visible on the horizon trailing a thin wisp of smoke from its truncated cone.

It was also explained that as Mars has only seven-degrees axis inclination, there is not much of a seasonal change. Water is very scarce on this arid planet; most of the precipitation falls at the poles. Vegetation is scarce too. There are a few varieties of prickly, cacti-looking plants. Also, near the polar ice caps grows a very luxuriant, green alga-like plant that follows the melting snow line. This spongy growth often attains a height, or depth, of forty to fifty feet. It appears and disappears with the season as it grows with tremendous rapidity, and it also disintegrates very fast.

There are also a number of species of lizards, reptiles, and some insects whose hard shells have enabled them to weather the extreme atmospheric conditions. Among them are giant ants that walk semierect on the two hind feet. The guide tells me these are mutants that were accidentally produced from a small ant in an atomic experiment ages ago. They are similar to humans in a very low state of intelligence, and at one time it became necessary to make war on them as they became so numerous and large. These strange ant creatures average two to four feet in height and live in rocky caves.

We did not tarry long on the surface. I followed my guide to a rather strange looking rock. Then, taking a small whistle from his coat pocket, he blew one note, and although I heard nothing, the rock immediately swung open disclosing a car-like elevator. We entered, and after the door closed, I had the familiar dropping sensation of our own modern elevators. The trip down took but a few seconds, and upon stopping, I stepped forth into what was my first glimpse of a Martian city.

I was immediately impressed by the soft white light that seemed to come from everywhere. We were standing near the entrance of a large tube. On Mars the cities are all underground and are connected together by huge oval metal tubes, which are from three to five hundred feet in diameter. There are monorail cars as long as our Pullman trains, which glide silently and very swiftly from one city to another. The bottoms of these immense tubes are used for parks, growing foodstuffs, and innumerable small manufacturing plants.

Because of the great distances between the cities, these tubes have been built only partially submerged. There are emergency air locks and bulkheads at the ends where they connect to the domes; other safety and precautionary measures are used to protect the cities and tunnels in case of breakdowns or outside attack. It is these tubes that have confused the astronomers on the earth. Some believe them to be canals, but there are also other theories.

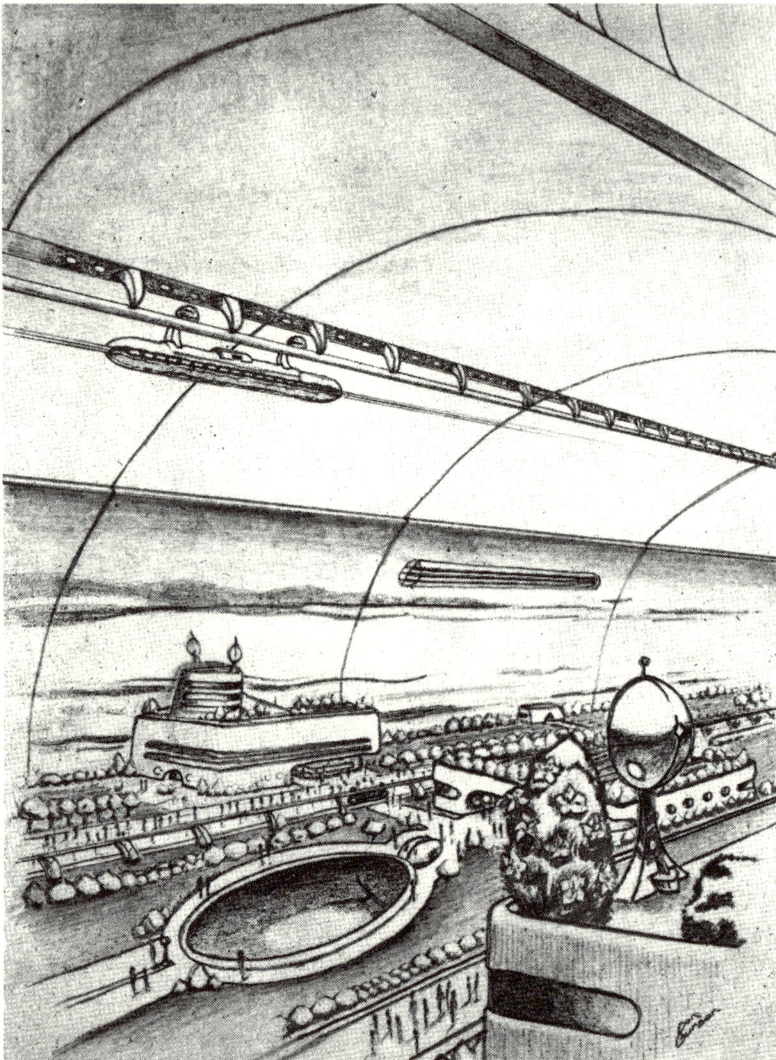

View of a street in one of the Martian underground cities, which are inside the metal tubes that have been constructed one mile below the surface of the planet.

The shifting desert sands often cover or uncover them, which leads to further confusion inasmuch as they seem to appear and disappear.

Turning about and looking down into the city is an unforgettable experience. Like all cities on Mars, it is built on the floor of a huge metal dome. These domes are sometimes four or five miles in diameter, and up to three thousand feet high. They are constructed of huge curved trusses of a whitish metal, seemingly of a magnesium compound.

These trusses are covered with a metal top and bottom, and the space between is filled with a plastic foam similar to the construction of the houses. This also gives added protection from the various cosmic rays as well as sealing in the precious air supply. Underneath the roof is an inner shell or a second false shell that is composed of sheets of pale blue plastic. This is suspended from brackets, from the dome, at a distance of about six feet; in this space are the many thousands of fluorescent tubes which make up the lighting system. They reflect downward the soft radiant light which I first noticed. I was told this light is very similar to a modified sunlight, and it is very healthful and stimulating to plant life as well as to the people. As these domes are built in the bottom of excavations, the sands soon drift over them and cover them up, giving added protection from the strong surface rays.

The Underground Cities of Mars

3

The Underground Cities of Mars

Martian cities are laid out like a wheel. The center hub is a very large circular structure that houses the various municipal and civic governmental departments. Underground is a very large atomic power plant for supplying the cities' needs. The streets stretch away from the hub like spokes, and at regular intervals the circular streets are intersected; this is similar to our national capital. The streets, which radiate from the hub, rise at a very gentle rate of inclination. The houses and other buildings are built on low elevations, which rise like tiers.

Walking up one of these streets gives one the impression of walking on air, as the paving is of a springy plastic material in a very soft shade of green. Stopping to inspect some of the houses, I am nearly overwhelmed by their wondrous beauty, simplicity, and charm. In every small, vacant space in the streets and grounds around the houses there are growing plants. These are mostly fruits and vegetables. They are planted in metal troughs and other containers. The soil is a mixture of natural and artificial plant humus and moss. The houses and buildings are semiprefabricated and come in a wide variety of plastics in pastel shades. The walls are formed of two sheets of thin plastic about two inches apart. After the walls are fastened together, a liquid, foam-like plastic material is injected or blown between the walls. After this hardens, it gives the whole structure tremendous strength. This hardened plastic foam acts also as a good insulator.

There is no problem of heat or cold in a Martian city with the abundance of atomic power. The whole city is air-conditioned, is free from dust and fumes, and is maintained at a constant temperature of about 68 degrees. Further comfort is assured the occupants of each home by its own air conditioner. Huge, electronic pumps suck in and filter the thin outside air and raise the pressure to about seven pounds per square inch. This is somewhat similar to the way in which a modern earth airliner pressurizes the passenger cabins. It also strengthens the overhead dome structure by pushing out uniformly at all points simultaneously.

As the outside atmosphere is very rare and of a low oxygen content, the Martian cities are becoming less and less dependent on that source of air supply. Many thousands of years ago they learned how to obtain air from water by electrolysis. They also make a great effort to create great underground reservoirs near the ice caps to drain off and store any surplus surface water, which also, along with the oxygen, has become increasingly rare through the centuries.

At the present time scientists on Mars are learning to make air and water synthetically, out of other elements. They have also explored every possible existing subterranean river or lake and have added much to the dwindling water supply by some important discoveries. It is estimated that—with careful conservation—they will have enough water for several thousand years, during which time other means will have been arrived at for solving this problem.

Mars is very rich in mineral deposits of which some, of the radioactive variety, are especially abundant. The planet has, also, like the earth, a molten liquid core of some of the more dense elements. Scientists on Mars are tapping these deep, fiery realms as an extra or added power supply, although the abundant radioactive elements are at present more than adequate to supply all the needs.

All buildings are supplied with electric power from the central power plant. The power is radiated over ultra-high frequency beams, which crisscross the streets, and is relayed by smaller substations. On top of each building is a split ball-like antenna that intercepts these power beams, bringing power down onto a small secondary radiator which, in turn, radiates the power through the building, lighting the lights, operating the various motors, etc. These are of course all constructed very differently than the motors and electric lights on Earth, which are large, clumsy, and very inefficient by comparison.

Window glass is a polarized material that transmits light one way, from outside in, which gives privacy without the problem of shade and drapes. A simple, metal, folding shutter is sometimes drawn across the window to shut out the light when sleeping, etc. Furniture in the home is very simple, it is contoured to the body, and is made of metal and plastic. The houses are not overly furnished as are so many of the earth homes, yet there is sufficient for comfort in a simple fashion. Rugs are of a plastic, foam-like material that is springy and resilient with no dusty nap. Various colors are used, as are slightly raised designs, which give variety and charm to the lovely overall appearance.

The kitchens would be a delight to the earth woman; all cooking is done in an oven that is built in a wall cabinet. The oven is operated on high frequencies, which cook all foods in a matter of a very few minutes or, in most cases, only a few seconds. There is a dial-like indicator, which is preset. The food to be cooked is placed in the oven, the door is closed, and a red light flashes. After a few seconds it is turned off, and presto, the meal is cooked. I did not see any boiling or frying in the conventional way. After dinner, the dishwashing is a very quick and simple process. The dishes are placed in a metal cabinet, a dial set, and after a few minutes all are clean and sparkling; no water is used. Instead, streams of electronic energy of some sort do the job. The dishes are made of a plastic-like material that is repellent to soil.

The bathroom is also quite different. The stool appears or disappears in the wall as needed. Disposal is efficiently taken care of by electrolysis. Very little water is apparently used. Bathing is done in a small booth where an atomized spray of pleasant-smelling liquid is sprayed on the body and wiped off with a very absorbent towel. Here also no water is used. There is some kind of energy ray used that stimulates and leaves the body very refreshed. Teeth are cleaned with a sort of electronic brush, which is a metal rod on a handle. Moving it around the teeth directs a flow of energy that cleans and stimulates the teeth and gums.

Very little of the normally expected house cleaning is done in a Martian home; all interior surfaces are dirt repellent, and as the air is normally very clean, and inasmuch as there is no smoking, frying, or similar soot producers, the homes are very clean and spotless.

Mounting a flight of stairs to the roof, we emerge onto a typical Martian garden. The roofs of these homes are flat and are planted with a wide variety of fruits, vegetables, and flowers. Each home grows quite a lot of the normal supply of foodstuff it consumes. They take great pride in these roof gardens and frequently engage in friendly competition in contests between neighbors in an attempt to raise the most beautiful displays of horticulture.

Returning to the street after my first glimpse into a Martian home, I became aware that there was something missing which was customarily found in a city. Glancing up and down the street, it suddenly occurred to me! There were no stores, no shops, no neon signs, nor any of the familiar garishness one finds in our cities. My bewilderment caused Nur El to smile. Motioning me to follow, he walked over to where a low, circular platform stood, up a few inches from the street level. We stepped on this platform and almost immediately began a slow descent through a metal tube. On coming to rest, a door opened, and we stepped out into what was another great surprise.

Now I understood why the streets slanted up and there was a lack of stores and commercial buildings, for here stretched out before me was a huge shopping center, which in some odd way reminded me of one of our annual state fairs. Up and down and around were streets and aisles with shops and booths displaying the many articles of clothing and food familiar to the Martian way of life. There were, however, noticeable differences: little or no advertising was used, and the shop owner sat or stood quietly by or worked on various articles he or she sold. A quaint system of barter and exchange is generally prevalent, although some form of script, currency similar to a department store charge card, is also used. Martians are inherently honest; stealing is almost unknown. Consequently, there is no need to accumulate more than is needed, for they do not have the fear of insecurity. There is no price haggling over various transactions. Some sections have mechanical, automat-like dispensers in which a keyed charge card is inserted and withdrawn after the article is discharged.

Another curious feature of these market places is the escalator sidewalks; on several of the main thoroughfares there were double tracks, one coming and one going, with a small bench-like seat to sit on. A person merely steps on, sits down, and is moved slowly up and down in front of the various stalls or shops.

Everywhere I turned to look I saw happy, smiling faces with none of that taut, drawn look that is so prevalent in our cities. I was informed that there are several of these "centers," which are regularly interspersed around the circumference of the cities and are located in the space formed by the floor of the dome and the rural sections above. Between these market centers much space is devoted to the cultivation of various crops, one of which is a grain very similar to millet. There is also a species of rice that grows with very little moisture. It seems that many of these plant crops have been evolved through a Burbank-like process to a point where they require the absolute minimum of water.

No heavy manufacturing is done in these cities, but there are several domes, which are devoted to and used almost exclusively for this type of work. But time was slipping by. Reluctantly, I followed my guide to another elevator, and we ascended back to the main level.

As we walked along one of the streets, I could not help but feel overawed by the beauty of all the things around me: the lovely homes and roof gardens, the landscaped parkways growing lush with fruits and flowers, the peace and quiet that was everywhere. I looked curiously at some of those who were passing by, although I do not believe they could see me in my astral state—at least, if they did so, they gave no indication that they were being rudely stared at—and I assumed their smiles and greetings were meant for Nur El. I did not see any indication of the use of any cosmetics on the women's faces. Their eyes were quite large and black with a distinct slant; their skin was wax-like and beautifully colored; their lips were red and well-shaped, which in all left nothing to be desired in any external adornment.

There are other things that were noticeable by their absence: there is no smoking, the use of tobacco being unheard of. Nur El chuckled when I asked about this, stating that such a practice was grown-up thumb sucking and was a habit belonging only to those who were not completely weaned. The drinking of alcoholic beverages is also unknown, so there were none of the familiar taverns or liquor houses. No doubt many of the eartheans would be lost on Mars without their familiar daily and hourly inhalations and libations of pap and bellywash.

The people of Mars are smaller than those on Earth, averaging only about four-feet six inches in height. They are somewhat Mongolian in appearance. The texture of the skin is very fine and soft, while the hair is usually straight, black, and quite fine. The men do not need to shave for they have eradicated, electronically, the growth of hair from their faces when still young.

Their clothing is simple with long, loose-flowing lines, with nothing to bind them, in many brilliant colors, including many shades we do not know of. All clothing is made of synthetic materials as no natural fibers are grown.

Martians are much older in soul-evolution than the Eartheans. They are a quiet, peace-loving people. Originally, they migrated in spacecraft to Mars from a dying planet. This was more than a million years ago. They also came to this earth and started a colony, but they found it impractical to maintain. Nur El explained that this colony became our Chinese race through the evolution of time.

Martian Science And Technology

4

Martian Science and Technology

Talking as we walked along, we soon came to the end of the street where it joined the center hub. Before us was a very imposing structure that formed part of the circular center section. As we entered I immediately became aware that it was some kind of laboratory. Various, strange chemical odors were faintly discernible. We were met at the door by another Martian named Yuzak. He was dressed in a bright yellow cloak or smock. Embroidered on the right side of the breast was a large, blue circle with a cross in the center, which denoted his rank as a head scientist.

After introductions he took us in tow, and we proceeded to make a tour of this great laboratory. The first section we entered was devoted to research, Yuzak explained. His science was very much advanced compared to that of Earth. On Mars the scientists know how to create and grow live tissue as well as plant chlorophyll. This is done by an elaborate process that involves the use of certain kinds of life-creating radiations or energy streams. It was explained that energy was actually, in a sense, tapped from the fourth dimension and that in such a state it was constructive and intelligent and could create life in artificial tissue or plasma. Passing on into another section, which proved to be a clinic, we found a number of doctors and attendants busily engaged in various activities. These people were dressed in bright blue garb. We did not tarry long here, but as we passed through, Yuzak explained that while there is very little sickness, the same knowledge of radiant energy was also used here to treat various physical and mental disorders. A malfunctioning kidney could be restored in a matter of minutes; or if damaged, it could be partially or completely regrown in a few hours.

Cancer and many other so-called incurable diseases are removed or corrected in the same electronic way. It was also explained that in all cases the patient was given a psychic diagnosis that correctly locates the true, originating cause as a psychic pressure or shock, incurred in either the present life or in some previous lifetime. This is extended psychosomatics.

Mental disorders, while rare, yield quickly to this treatment that quickly removes or rectifies these malformed vortices or thought wave patterns, which have been incurred in the subconscious or psychic body. There are no jails or prisons; crime is considered a mental disorder and treated thusly. Such treatment is kept secret and not exploited as we do; therefore, there is no deflation of the ego. There is no pain or shock or lengthy doctoring. Usually the patient is home in a matter of hours.

But here again we have come to another section. In the rear of this great building, upon entering, I am again impressed with the tremendous array of strange and unusual looking apparatus.

I have visited laboratories but nothing like this! Going over to the side wall where there is a row of small glass tanks, something like aquariums, I noted that over and around each one is a tremendous array of tubes, pumps, cables, etc. Looking closely into the tank, I see a yellowish, cloudy liquid with something in the center that looks like a baby! Recovering from my surprise and shock, I was assured by Yuzak that it was a baby. He went on to say that their science had advanced to such a point that this laboratory development was possible under conditions similar to those of natural growth in the womb of the mother.*

First, a capillaric system is grown from two main tubes, spaced about a foot apart, and a web of tiny tubes form. These in turn unite in the center and form a structure similar to that found in the body. This structure is grown in this aquarium-like tank in a plasmic-like liquid under radiant energies. After the capillaric web is formed, a membranous tissue grows over all. This is the artificial

* *See Appendix A. Japanese Scientist Develops Artificial Womb*

View of a portion of one of the Martian laboratories that serve as an artificial gestation chamber used to bring a baby into existence when natural birth is impossible.

placenta. The ovum from the mother is then attached and inseminated with the father's spermatozoa. From then on, the fetus grows in a normal way, being attached to the placenta with the umbilical cord supplied with the natural blood supply, which actually, in part, comes from the mother. Under these controlled conditions the natural incubation period may be reduced by one-half. The blood supply is constantly refreshed and purified from a central blood bank or with such similar Rh factors.

Nur El further explained that this laboratory process was only taken advantage of by those women who could not ordinarily bear a child, or where there were other and sufficient reasons for so doing. Passing through this laboratory, the guide went further into the problems of birth control and sex. He stated that children are usually limited to two per family so as to prevent overcrowding and a lessening of an advanced family relationship. The ratio of births is usually determined by the death rate average.

Sex relations are very sacred and considered a great creative gift. There are none of the usual sexual stimulants such as advertising and spiced, heavy foods, etc., which tend to overexcite the people of Earth. Consequently, sex assumes its rightful place in the life of the Martians.

These people have a basic spiritual concept that teaches them from birth the importance of loving one another and finding their greatest joys of life in doing for each other, not doing to each other. I did not see any of our familiar churches and steeples; it was explained to me that there are none. Worship is not a pagan-like bowing down to some mythical god, or gods, but a twice daily observance to the Great Infinite Creative Source, and there is a once-weekly community observance giving thanks to this Source.

Other spiritual aspects of Martian life include communications with those who have passed into the spirit world; in fact, every Martian considers his spirit friends and relatives a part of his daily life. No doubt our Chinese have derived their ancestor worship from this source.

The Martians have also developed reincarnation to a point where it forms an integral part of their lives; they plan for a future time when they will relive a new life among old friends and relatives. Many children frequently identify themselves as former loved ones.

I was tremendously impressed and overawed by all that I had just seen, when, passing out of the building, we entered the courtyard or foyer to the central section which is a large community stage and theater. Sitting there to rest, my guide went on to further explain some of these fascinating facets of life in these cities. He was still talking of children and said that soon after the birth of a child, all the potential mental faculties and quotients were determined by an electronic diagnosis, and any criminal or negative characteristics were removed by the radiant energy process. The child was further conditioned against such recurrence. There are no public schools. The child is taught to a large extent in his sleep by a "Z-ray," which imparts the lessons or knowledge directly into his subconscious mind. This ray can be likened somewhat to a radio frequency which carries the spoken word, yet it is inaudible to the ear; in this case, however, it is received and stored for use in the child's mind.

Usually a child will have the equivalent of a college education by the time he has reached the age of ten. Such schooling is done to bring out the best points of character and to especially train him in whatever vocation he is best suited for.

As I listened to the soft, accented voice of my Martian friend, my mind inadvertently began to recall and compare scenes of my earth life—things like the roaring streets and highways, the stench and smell of thousands of cars, the hate and greed and avarice— with the simple, quiet way I was just beginning to glimpse and understand. Nur El caught my thoughts and for a moment stopped speaking. A slow smile lifted the corners of his mouth and his eyes began to twinkle. "No," he said, "these people would not migrate to the earth. First, they would have to become accustomed to the difference in air pressure, and if this were done suddenly, it might be very dangerous, like a diver going down beneath the water too quickly." He paused a moment, then continued: "Then there would be deadly disease germs and the many viruses that we here on Mars, not having had such things for thousands of years, have lost our resistance to."

I could see his point, but I wondered a bit as to how they knew so much about the earth. Patiently, he explained that there were semi-surface observatories with electronic telescopes as well as a variety of radio and radar-like devices which gave them a very good idea of what went on there. Besides some of the more advanced scientists were masters at astral flight. Even an ordinary citizen of Mars was quite adept at mental telepathy, and this type of communication was used as much as speech.

But I was curious about the large theater that was confronting me; however, I waited until the explanation came that this was the large, central theater or community playhouse. Here the various plays, pageants, and observances took place. There were also many civic orchestras, and various kinds of folk dances were given here. Many of these customs are of ancient origin and the meaning of some has almost been lost in antiquity. I was privileged to see part of a spring festival. Here on the great central stage several hundred children were gathered in large circles all decorated profusely with flower garlands; in the center was the queen of spring. The children were all moving slowly to and fro and chanting a slow singsong rhythm. It was all very beautiful and colorful.

Going back into the ring-shaped building, we emerged into what were some of the chambers used for judging or administrative phases. The government is of a very simple form. I was amazed when told there are no written laws. Each citizen lives under a simple understanding, an unwritten code. It was a very reasonable facsimile, if not the actual golden rule. In other words, do for others first. If a person acts selfishly, begins to steal, or shows symptoms of anger, he is considered ill and treatment is quickly administered.

Each five families have a group leader or "Icla," as he is called. He represents this group and is responsible for their general welfare. The judges or heads of different departments are chosen on their merit, and it is usually done through elimination, examinations which require a lifetime of special training. There are no political systems. Brains and character alone determine a candidate's fitness for an office. Male and female are regarded as equals, with no discrimination shown. There are no old-age institutions in these Martian cities. Great respect is shown the aged, and they live with their children until the time of passing. No doubt the Chinese on the earth brought this custom of respect and veneration down through the ages from their Martian ancestors.

Passing along through a hallway, we again emerged into what I immediately gathered to be a library. Here a number of persons were busily engaged in sorting and caring for a large variety of books—if they can be compared to ours. These were, for the most part, rather ancient scroll-like volumes, but there were others that resembled, somewhat, our own bound volumes. I was told that this library was kept mostly for research and for historical reasons, as comparatively little reading or writing is done in our usual manner. Most records are kept in memory machines or some other unfamiliar, electronic device. The alphabet consists of a number of cryptograms. I did not linger long enough, however, to go into things of this nature in detail. Another thing that im-

The monorail transportation system of the Martian underground cities.

pressed me was the wide variety of pictures and objects of art, which were everywhere. These people are exceedingly artistic, and almost everyone spends some time at his particular chosen expression, decorating screens, ceramics, furniture, etc.; all were given some treatment whenever an opportunity presented itself. It was all in very good taste, however, and most pleasing to the eye. These traits are quite evident in our modern Chinese.

Going through several of these offices we again emerged into the open air. Before us was one of the larger radial streets; coming down this street was what looked like a silver gondola of some sort, suspended from an overhead rail. Going closer to examine this strange craft, I found that it was about twenty feet long. It had six or eight bucket-like seats. There is a rather elaborate system of control used. All is done, of course, electronically. There is an eye on each end to keep it spaced at a reasonable distance from other cars. Gyroscopes are used to prevent sway, and it is powered by a motor in each flanged wheel, which rides a single rail, suspended at short intervals by metal standards. These cars are stopped with a single blast of a noiseless, supersonic whistle and started after the passengers sit down. A number of the main radial streets have this monorail shuttle car system. Others used moving sidewalks, somewhat similar to escalators.

The Martian Society

5

The Martian Society

At this point my inspection of these very interesting monorail cars was interrupted by what appeared to be a gay group of people going into one of the nearby homes. As they disappeared, laughing and chatting through the doorway, I turned and looked inquiringly at Nur El. He smiled as he informed me there was a marriage going on. He went on to explain that marriage is very sacred here; very rarely is there a separation or divorce. As children, couples are selected according to their mental quotient and their compatibility. Further work in preparation is done until the time when they are actually joined in marriage, usually around the age of twelve or fifteen, as compared to our years.

The marriage ceremony lasts for about three days. The first day is spent in the exchanging of gifts between the respective families. The second day is devoted to mind projection and various semisecret preparations. The whole affair climaxes on the evening of the third day with a banquet. Sitting on opposite sides of a long, low table, the close relatives of the bride and groom eat a silent meal, not glancing up or looking across the table during the entire time. This silence is broken by the two fathers, or in case of their absence, the mothers can substitute by the sharing of a cup of juices made from certain fruits. After this, the cup is passed around and the whole thing ends up with several hours of merrymaking. Then a parade takes the newlyweds to their new home.

Returning to the subject of transportation, it seems that these people do not travel much as compared to our earth people. Of course, there is considerable inner-city travel over the monorail car system, which I saw in the tube. There is however a kind of communication that renders a great deal of travel unnecessary. This is a form of telephone, or teleview, as it could more properly be called.

Besides conversing, each party can see the other one through a small screen similar to our television. This, of course, can be shut off by manipulating a button in the event there is a need for privacy. Speaking of television, their system is far more advanced than ours; the screen is built into the wall of the room and is about four by five feet square. All programs are in three-dimensional color, very lifelike and natural. Such programs are, of course, the very highest type. There is only one channel to a city, and all program material is produced and telecast by the people themselves. There is no advertising or sex intimations but only such things as the festivals, lectures, various stage presentations or musicals, which take place in the central theater. Little or no news is broadcast, and then, only that which is of a nature that would not cause fear or restlessness.

Music plays an important part in life here. Most of the instruments are of the string type and are usually plucked. There are some reed or flutelike instruments, which help give variety. There are none of the heavy percussive types, which form a large part of our modern orchestras. The music itself is, for the most part, a quaint sing-song-like rhythm or chant, which usually depicts some story or moral lesson—even historical events are portrayed. Considerable color is used in the stage presentations, which gives much added charm as the innumerable color combinations rise and fall with the rhythm of the chant.

Since my initial trip, I have returned to Mars several times and have learned much more about this fascinating civilization. To those who are proponents and ardent supporters of our free-enterprise system, let it be said that they have a great shock coming to them. On Mars there is no dog-eat-dog competition such as we, on Earth, are so familiar with. Everyone works for the government because the government is the people! This highly-developed, socialistic system is not to be confused with any so-called communistic governments on Earth. The Martians never break laws, consequently, there are no laws.

They have long ago eliminated legislative bodies. How different here! We have a huge and vast intricate network of legislative bodies, as well as various branches of law enforcement. The average American has many thousands of laws to obey. The great majority of the people either knowingly or unknowingly are breaking laws. As fast as a way is found to circumvent one law, a new one is passed to prevent this!

The modern Chinese have placed a great deal of personal value on face; they would rather die than lose face. This sense of personal integrity was brought down through the ages from their Martian ancestors. The average Martian has an advanced state of conscious and personal integrity. This eliminates the ponderous and very expensive system of government to which we are accustomed. There is in consequence a vastly simplified way of life. There are no taxes, as this land is run like a highly-ordered, nonprofit business. The various departmental or executive heads are all highly trained specialists, and they hold their positions because of ability and integrity.

Another thing, which some earth people would miss on Mars, is money. They actually use no money. Therefore, there is no banking system. The citizen of Mars is, in a sense, his own bank, and his medium of exchange is his honesty, his credit. Everyone works; there are no loafers or unemployed. They all take great pride in their vocation. Work is not done for personal gain or selfish reasons.

There is no sense of insecurity on Mars for all are provided for, including the aged and the incompetent. A working person is provided with a metal plate whereon are stamped his or her name and an identifying number. This plate is used in the large, automat-like dispensers, which I was shown on the lower floor level. When a person desires food, clothing, or whatever the need, he merely inserts this charge card, if I can call it such, and withdraws it along with the article. As he is a very honest person, he never gets more than he needs or is entitled to. This is done strictly on his own personal sense of honor and integrity. As he has no sense of insecurity, he never hoards or takes more than he requires. We can imagine what would happen if our great hordes of dishonest, grasping, selfish people were turned loose in the Martian cities!

There are none of the conveyer belt techniques that we use here, although they are very advanced in science, electronics, and machinery; yet, there are a great many handmade products. They realize that a vast productive system would only defeat their way of life. The more a man works the more he can take home, so if a machine takes his place, he has nothing! An article is valued only for its usefulness and not for the profit it might bring. There is some semiprivate enterprise. In case a person makes some product in his spare time, he may open a booth where it is displayed. All such enterprises operate on a barter exchange system. As there are no profits, these private exchanges consequently are not running in competition with any other.

I have gone to some length in trying to point out and explain what is very obviously a superior way of life.

We Americans have for a long time prided ourselves as having the best way of life on Earth. This is quite true as far as Earth is concerned, but not in a broad, unlimited view. As Einstein once said, "We are never fools until we quit seeking."

In our present democratic system, which is actually capitalistic, we have long passed the point of diminishing returns. A government should be of the people, and we should not lose this goal in our haste for profits. The average citizen of the earth has a state of existence something like a whipped circus animal. Yes, it is quite true that we have a higher level of life than many other countries, but what a price we are paying for this! And could not we get something better if we tried? But, back to our Martian friends.

From childhood they are taught to be useful and productive. Because of their simplified living habits, they have more time for self-improvement and for developing new types of plastics, textiles, etc. They usually limit their meals to only two a day and even those are very simple, consisting primarily of vegetables and fruits, with some synthetic foods. The Martians are not meat eaters, partly for the reason that animal life has largely passed from existence on Mars, with the exception of mutants and the few obscure species previously mentioned, and these are inedible. The Martian has learned how to grow and also to synthesize many spices and to produce, artificially, many protein foods. Many nuts, some of which look somewhat like peanuts, are raised. A large nut similar to a pecan is grown on vines. There is a vegetable something like a cross between cabbage and lettuce, the leaves being three-fourths of an inch thick and about a foot wide! It is light green in color and very nourishing. Another common vegetable looks like our onion, but it has large, flat, flared leaves. The people produce artificial sunlight in huge tubes so they can grow foods in their underground cities.

Space is not too plentiful. There are some dwarf trees, three to four feet high with a fruit that looks like an orange, but it has meat like that of an apple, is red skinned and sweet. It is called sit-yu. Some other vegetables are grown in troughs in long rows. These look like huge mushrooms and have a delicious meat.

Science & Philosophy
The Harmony of Martian Life

6

Science and Philosophy
The Harmony of Martian Life

Martians are closely connected with the Venusians through thought transference or mental telepathy. At one time interplanetary travel was used, but these spacecraft are at present stored in huge underground hangars. They are being held in readiness for any emergency, such as a sudden mass evacuation of the population should any unexpected need arise.

Yuzak explained that the science of interplanetary travel was something very difficult for the earthman to understand, inasmuch as the people and the craft itself would, in taking off, actually change the rate of vibration of their own and the craft's atomic structure. Thus they would in a sense become weightless and temporarily free of various usual forces, such as gravitation, inertia, etc. This is something like the preconditioning for fire walking, which is still practiced in different parts of the globe.

My guide went on to explain that this science of changing the atomic vibration rates is a very advanced one, and that if the earthmen would learn this, it would remove all the present-day obstructions and barriers, such as materials, fuel, pressures, and the hundreds of other hindrances to present-day space travel. At present their experiments have progressed to a point where a man's body can be changed into electrical energies, sent over a radar beam, and then changed back to the original state—all in a matter of a split second, and with no pain or discomfort to the individual.

This no doubt sounds fantastic, but remember that this same thing is being done here frequently, from the spirit side of life. The memoirs of Sir William Crookes give an example of this in his experience with Katy, who materialized after being "dead" for several years. She was so real that he embraced and kissed her. Afterward he said she was as real as his living wife! Nowadays, it is only the ignorant man who scoffs at any new ideas. No doubt there will be those who will be unable to believe these truths, but it does not matter, since that still does not alter the facts. Just fifty years ago they were laughing at the Wright brothers—and think of what has been invented and improved and brought into use since that time. This is no less so in the future. The things of science that are commonplace in the Martian way of life could well become a part of ours in some future day.

I was informed that the Martians understand what

is happening to the earth people and its veritable rat race, and they are very desirous of aiding through mental telepathy to inspire as many of the eartheans as are receptive to their ways for the advancement of mankind and the improvement of conditions on this planet, and within man himself. Until man realizes the great overbalance on the material side and gains the necessary spiritual knowledge, we cannot hope to be harmonious with the other more advanced planets. They say that it is realized only too well that they must not interfere with the evolution or progress of man on Earth to any great extent, for it would not be in keeping with God's immutable law of individual soul progress.

The scientists of Mars have informed me that our telescopes, in the photographing process, are subject to error and do not get refraction but infraction. Sometimes the light rays, or vibrations as they are more properly called, are at times subjected to distortions, or bendings, in their flight through space due to the proximity of some other planetary body. Because of the conjunction of magnetic lines of force, the astronomer does not always get a true picture of what he thinks he is seeing. Also the planet Mars, on the outside of its surface, has a tremendously-charged shell. While it is invisible to the eye, this can cause great distortions in light-ray frequencies.

And so the time has come, at least for the present, to return to our more familiar planet Earth, and I do so reluctantly, for there are still many more facets of life on

Mars—rather in Mars—of which I have only a slight understanding. I would like to learn more of their various customs, celebrations, and observances. Obviously, it cannot all be taken in during only a few trips. I cannot recall a single instance of this most fascinating and interesting experience of my visits to Mars without a deep feeling of awe, reverence, and gratitude to this very fine person, who calls himself Nur El, for giving so unselfishly of his time and efforts in explaining so many things about the planet, its people, manners, etc.

But the most important part he is playing, along with many others on their planet—for all have this one thing in common—is to present the plea for man to heed their warnings and put the use of atomic energy to constructive use, instead of in bombs. Through their ability of clairvoyance and mental telepathy they can see what is happening, and they are tremendously concerned regarding their own planets For anyone who understands even the slightest degree regarding the fourth dimension can't help but realize that the destruction caused by the explosions and other results on the physical plane are indeed a very small part of the upheaval that takes place in the higher realms of existence, and even on the other planets. When throwing a pebble into a pool, we can only see the ripples with the physical eyes, but still down, on and on, goes the stone regardless of whether we see it or not. The men on other planets are pleading, for their own sake as well as ours, for us to cease this nonsensical, destructive thing that can only lead to tremendous repercussions in the ethers, even to a far greater degree than in the material or physical way.

There is also a grave concern by these people about our destructive, downhill way of life. We are creating and breeding a race of psychopathic misfits in our highly-specialized, mechanical world. The people are becoming robots. They cannot sleep for nightmares from fear and insecurity. Their days are an endless succession of almost frantic scurrying, or worse, a robot-like existence of work, sleep, and work. On every hand, flagrant psychological and sexual stimulants are used for advertisements; suggestion is used to hammer home these cheap and malicious messages until the brain becomes numb and neurotic. Exploitation of the masses has become a highly specialized science, ruthless and cold-blooded, running the gamut from charity to vice.

On the other hand, there are almost equally frantic attempts being made by various religious groups, churches, individuals, etc., to portray in some way to the great masses the grave dangers confronting them. These efforts are pitifully small and weak. Moreover, these efforts are not above suspicion, for there are here, as elsewhere, many charlatans.

The Supernova Connection

7

The Supernova Connection

Resolving all these facts, and in trying to establish and correlate an active, vital, spiritual concept in their hearts and minds, earthlings must find their own solution. Quite likely, in some future day, man will evolve a way of life that will embody not only some of the old spiritual and psychological facts that have been tried and tested, but many new and advanced facts will be added—new at least as far as the majority of the people are concerned.

In this spiritually and scientifically-integrated philosophy of life, man will find not only surcease from this present-day's mad race to oblivion, but he will find instead a new concept of the continuity of life, and with a Garden of Eden life will come happiness, creativeness, and fearlessness.

It was intimated in the foregoing pages that it was

quite obvious that it would require several trips to gain a comprehensive understanding of the Martian way of life, and that therefore I would quite likely, from time to time, make similar visits or flights. Since concluding the last pages, Nur El has made contact several times and has given or shown me pertinent information regarding several issues that I would like to clear up.

In case some of you are wondering, just as I did, what caused his people to go underground and why they have not migrated to some other planet, since they have all the necessary craft to do so, Nur El explained all this by first saying that Mars was, up until about 100,000 years ago, a planet very similar to the earth. There was air, water, and an abundance of plant and animal life. The cities flourished on the surface just as ours do. At that time, through their interdimensional science and also with their superior telescopes, they saw, somewhere out in space, a cataclysm take place. One of the giant suns suddenly went berserk, flared up like a nova, and then exploded in a terrific, blinding flash. Huge chunks hurtled out into space in different directions, each one a smaller, white-hot, atomically-burning sun, shooting off great streamers of atomic energies. It was determined by calculations that one of these fiery pieces would pass very close to our solar system. As it was larger than our own sun, it was conceivable that there would be tremendous repercussions; in fact, anything was possible.

Because of the great size, it would have a tremendous gravitational pull, besides giving off great energies.

It was also determined, as light traveled much faster in space than this huge chunk, that it would pass our solar system. This, therefore, gave the Martians a grace period for preparation. They had, however, a choice: to stay on Mars or to migrate to another planet far away from this solar system. After a search of the nearby, practical limits of the heavens, it was found that there was no other planet available which would be suitable. So an alternative was decided upon. They could build huge cities underground! During the next two hundred years or so, an almost frantic building program was assiduously pursued. We can well imagine some of the problems, the sacrifices, and the labor and research that went into this tremendous project, but it was finally accomplished. In due time all was snug and shipshape as possible. Buried deep under many feet of rock and earth, in their newly constructed domelike cities, the people of Mars waited for the final hours. Day by day they watched the white glowing mass of light grow larger and larger. There was no really accurate way by which to measure how close it would pass or just how hot it would be. No doubt many thought it would be the end!

Finally the hour struck. Nur El stated that for eight days the planet was rocked, torn and twisted, as great forces blasted and ripped the surface. Plant and animal life disappeared almost entirely, except for the few species that escaped by being buried or by being in caves. The air and water, too, were largely dissipated or drawn off with the passing nova. When it was over, Mars was decimated and burned to a cinder.

Other planets also suffered. On Earth there were great earthquakes and tidal waves. A great continent and civilization, called Lemuria, sank beneath the sea. Great deserts were burned into the surface in places that were formerly beautiful forests and plains. The Sahara and Gobi deserts were two of these. Earth's orbit and axis, or the poles, were also changed. Instead of a circular orbit Earth was rocked into an elliptical orbit, with the addition of a very slight but definite wobble or oscillation, which it has never lost. Uranus was pushed out into an orbit much further away from the sun. Neptune suffered a similar fate although not quite so pronounced.

It is conceivable that the other planets were also affected according to their size and position at the time of the passing of this huge celestial "atom bomb." No doubt the knowledge of this cataclysm will clear up some of the mysteries of Earth's history, which have been puzzling the seekers of truth for many years.

Another question might be asked in regard to the Martians' system of time. As nearly as can be determined, they maintain a daylight and dark cycle of days just as we do; this is done by simply dimming the overhead domelights at regular intervals. A week is about twelve days. There are apparently no months tabulated, but the year is divided into four quarters, something like seasons. Their sense of time is quite different from ours. They practice none of the hurry-scurry, which is so prevalent here. Thus, there is no constant clock watching, or the tensions which such a practice produces. Here again is a similarity to the Chinese, i.e., "There is always tomorrow," which is a familiar Chinese adage. Not that it should be construed that they are a lazy or indolent race, far from it. They are very industrious, but they mark each day by their accomplishments and not with the passing of hours.

Just in the event that there may be individuals who may take personal resentment, or a wrong view of any comparisons which have been made as to forms of government, etc., for the record let it be said: I consider myself just as much a loyal, staunch American as any other citizen. My parents and grandparents were all pioneers and helped carve out of a wilderness what is now known as the state of Utah. Moreover, my views on present-day political systems are comparatively average. As a philosopher and scientist, however, I might point out, as others have, many obvious flaws, which are a matter of present day concern. It is the task of our legislative bodies to anticipate and to circumvent such flaws. The constitution is, in its entirety, a very wonderful document that expresses very highly-evolved concepts of living. Such documents and governments resulting from them are, however, effective to the relative degree in which the so governed people integrate them. The end result, therefore, devolves upon each individual so participating. The fact that there are twenty-one amendments proves the correctness of this reasoning.

We must always remember that at no time in life should we ever assume we have reached perfection. We must always visualize an infinite universe in which perfection is only relative. We must not become restless or dissatisfied with what we have. Rather, let us be thankful to our Creator for each daily manifestation of his love and abundance, always secure in the knowledge that he will always so manifest. Our forefathers, and those who helped carve out this great nation, had such idealism and faith. Because we have countless conveniences, which they knew not, we must not lose our initiative to keep pushing back the frontiers of ignorance, fear, and insecurity.

Our present civilization, good as it is in some respects, leaves much to be desired. It has been the purpose of Nur El, the people of Mars, and myself to bring you some understanding of their ways of life, hoping thus to bring about, among the nations of the world, future interplanetary relationships that are harmonious and conducive to a better way of life.

We all salute you and we wish you infinite love, wisdom, and peace.

Addendum

Addendum

Shortly after writing the manuscript of *The Truth About Mars*, there appeared in the *Saturday Review,* on May 28, 1955, an article written by Dr. Robert S. Richardson, astronomer at Mount Palomar. This article refers to the findings of the International Committee on Mars, which closed its fourth conference on March 25, 1955. There are several interesting statements made, on which I will comment. Mr. Richardson states that the consensus of opinion is that there is life on Mars, or that it could exist. This is at least some progress in the right direction. Also, he is quite correct about the deserts. Most of the surface of this planet is semiarid wastelands. It is however incorrect to say there is no oxygen there.

The presence of the green maria, which he calls sponge-like alga, proves the presence of oxygen, although in a comparatively rare state. As everyone knows, oxygen is necessary in the breathing cycle of any plant that contains chlorophyll (oxygen on Mars is about 10% of the density of the earth's oxygen) as was stated in the aforementioned article. This sponge-like algae is found growing along the edges of the snow banks, and it often attains a height of forty to fifty feet. It dries up with the vanishing of the snowcaps and regrows the following spring.

What the doctor states regarding water on Mars is very true. It is very scarce, and the people of the underground cities take great pains to conserve every gallon of it.

As to the temperatures, that is still a matter of conjecture. It is a very tricky business to measure heat over a distance of thirty-five million miles. Moreover, surface temperatures do not affect the inhabitants of the underground cities, as they are completely pressurized and conditioned with temperatures maintained at a comfortable level.

The most surprising part of the entire article was that there was absolutely no mention of the famous canals of Mars. I wonder what happened to them? These canals were for many years a great controversial subject. The photograph in this article does not show them, but this is perhaps explained by the fact that this photo was taken with infrared light film. It may be that the savants at Mt. Palomar would like to explain this.

It is of questionable value that the writer interjected such a material angle as real estate. It seems it would have been wise to confine the remarks within the domain of science and leave this problem up to some of our great promoters in that future day of landing. It might also be that the Martians would resent our tearing up their planet.

It is also probable that in that future day when man does have space travel to Mars, he will be able to take his wife or loved ones along. Landing there will be somewhat like taking a plane to a far-off city on this planet. Upon arriving on Mars, the spaceship will be taxied into a huge airlock. The passengers will disembark and find hotels and accommodations in a similar fashion as on Earth—assuming of course that such factors as freedom from germ life, health, adaptability to lower air pressures, etc., have been fully compensated for.

It is unfortunate indeed that the astronomers of today take such a dim view of the possibility of life on other planets. They should be in a position to know better than anyone else. Does it not seem a bit preposterous to assume that in all the countless billions of suns, star clusters, galaxies, etc., and their associated planetary systems, that Earth and Mars alone are inhabited?

During the last year or so, 1966-1967 of my earth mission, I have frequently paused and have gone back into the different books and other literature of Unarius that have been brought forth since the actual beginning of this work in 1954, and I have found one work that is perhaps most interesting, and which has proven very prophetic, to be the book *The Truth About Mars*. This book is, incidentally, the only one which I personally, laboriously wrote in long hand.

Since writing these pages, many wondrous happenings and discoveries have occurred. At least one or more unmanned satellites have been flown to or around Mars, or have actually landed on its surface and have sent back hundreds of photographs as well as information on other planetary aspects, all of which have only further proven certain surface descriptions which I gave in the book.

But until the first spaceship carrying people from earth actually lands on Mars, the earth people have yet to factually find out about the many underground Martian cities and the highly scientific and completely socialized way of life enjoyed by the Martians.

Back here on Earth, however, much has been discovered which has duplicated, unknowingly, the many scientific accouterments of life used by the Martians. Cooking by high-frequency rays is an established fact in this time. Quite recently two American scientists created an artificial placenta and actually grew tissue, and in the article there was great emphasis on the future time when babies would be grown under the same conditions.

In many hotels, public places, and homes there are walls and ceilings made of plastic material that glows brightly when a button is pushed, just as the walls and ceilings of a Martian home. Punched cards are used by computers and processing machines, and also they are used instead of keys for cars. Great progress has been made in wonderful new and synthetic materials. Styrofoam is now being used not only to package articles in shipment but for insulation and in actual construction, along with fiberglass and other synthetic materials. Houses, boats, car bodies, and aircraft are but a few of the many thousands of different achievements in plastics that have been accomplished since the first printing of the Mars book.

But there is no doubt that the greatest single accomplishment was, or is being done by Walt Disney—now deceased—and Disney Enterprises.

Shortly after the publication, by about two years, Disney Studios brought out a short movie about Mars in which were depicted the huge, domed, underground cities and connecting tunnels, all done as a colored cartoon feature. Shortly afterward, Disney actually produced, as an exhibit in Disneyland, a Martian home. Could Walt Disney have obtained and read *The Truth About Mars*?

During the 1950s, books were sent by two, or more, Unariun students to many prominent and well-known personalities in the United States; yet, at the moment, it is not known whether one was mailed to Walt Disney or not but very likely so. Also, these books were placed in the public libraries immediately upon their publication. Mention might be made here, too, that these first Unarius books *The Truth About Mars* and *The Voice of Venus* were also sent to the President of the United States as well as to the Governor of California, Goodwin Knight, whose wife acknowledged their receipt.

During the early months of 1967, Disney Enterprises announced they were building an amusement park in Florida that would be housed under a huge dome of transparent plastic and metal, and it would cover about 80 acres. Within this dome, in a totally controlled, conditioned environment, there would be not only the amusement park, but it would also provide housing and homes, hotels, etc., for at least 40,000 people. The picture of this proposed amusement center and its dome has appeared in color as a nationally syndicated feature in different magazines and newspapers throughout the country. The picture clearly depicts the central, radial plant and structure with its central hub as seen and described in Martian cities.

Disney Enterprises also announced there would be a number of similar projects, even much larger, which in the future would be constructed elsewhere in the United States and in foreign countries.

On the heels of this announcement a very famous architect living in New York stated and advocated that in some not too distant future all large American cities would be housed under these huge domes! For example, New York City, so enclosed in this totally conditioned environment, could pay for all the construction costs of this dome, within ten years' time, with the money saved from snow removal in the winter! Doors and windows in all buildings could be left open the year round in a balmy 78-degree temperature. There would be no dust and no smog; automobile transportation would be entirely eliminated in these cities and supplanted by electrically powered vehicles and other means of transportation, such as moving sidewalks, etc.

It would be well to note that here in Southern California there have been two malls or shopping centers, recently constructed, which use moving sidewalks; and there are several blocks of landscaped parkway between the rows of shops where no cars are allowed, just as in the Martian cities.

At the time of this writing, it is indeed a prophetic time. The most recent satellite of the lunar probe series has just landed on the moon, and within the next five years it is quite possible and highly probable that the first man will be walking on the lunar surface. How soon, then, will the first spaceship land on Mars, and what will the meeting be like between the Martians and the earth people? As we speculate, we can only hope the earth people will be sufficiently advanced by that time to be compatible with the Martians, and that these earth visitations will not contaminate this wonderful, highly-developed civilization with all of the viruses and germs, all of the diseases and decadence, the ever-increasing immorality of this present earth time.

We do not need to conjecture or speculate on any probable future time in earth's history when a physical earthman would set foot upon any of the seven celestial planets described in the Pulse of Creation series. These planets and the people who inhabit them are not in the physical, material spectrum of this earth. Even Venus, the mystery planet, if ever contacted in reality by an earthman, would be vastly different than he has yet surmised.

Meanwhile, as I sit in my easy chair and watch the passing of days, I note the many "discoveries" and achievements made by science as time progresses, and I chuckle a bit; I even feel elated, for all of these things and many more "undiscovered" by the earthman have been presented in the Unarius library. Yes, and if I lived in this manner in this body for the next thousand years, I would see no end to these discoveries and revelations, all so presented in our books and lessons during the eventful years of the 1950s.

During the late months of 1966, two prominent and well-known scientists at the Berkeley University made an important discovery. Using the satellite *Imp*, they mapped the magnetic lines of the solar system. It should be noted that these same magnetic structures, similarly diagrammed, were presented in the second lesson course of Unarius, printed and copyrighted in 1960! This is one of the many daily happenings and "discoveries" which cause me to chuckle and smile and even to indulge in some "I told you so!"

(Note - June 20, 1967: At this later date, two biological scientists have kept a human fetus alive and growing for a period of 58 days in an artificial placenta.)

Appendices

Appendix A

Japanese Scientist Develops Artificial Womb

TOKYO — A Japanese scientist has developed an artificial womb capable of incubating goat fetuses, but it may take 10 years before it could be used to save human babies, he said on Friday.

"We're aiming eventually to use the technology for human fetuses, but it will take maybe 10 years," Juntendo University gynaecology professor Yoshinori Kuwabara told Reuters.

Kuwabara's findings have been published in the *Journal of the Japan Medical Association*, the country's medical review publication.

Kuwabara said he had maintained goat fetuses for up to three weeks in a plastic tank, until the end of their incubation period, but it would be some time before it could be used for human fetuses.

"Other researchers around the world have managed to maintain fetuses outside the womb for up to several days, but we're the only ones who have sustained life in an artificial womb for so long," Kuwabara said.

The artificial womb is a rectangular clear plastic box filled with amniotic fluid at body temperature and connected to an array of vital functions devices.

The fetus lies submerged in the tank womb which replaces vital oxygen and cleans the fetus' blood with a dialysis machine connected to the umbilical cord, he added.

Juntendo professor Koyo Yoshida, a member of Kuwabara's research team, said he was wary of exaggerations by the press of a miracle artificial womb which could free women of the pain of childbirth.

"Some of the stories printed in the British press are only half true. We are not trying to extend the possible incubation period in the artificial womb right now. Three weeks is the limit with the technology we have now," he said.

Many of the experimental goat fetuses incubated in the artificial womb only survived for a few days after being taken out of the tank although some remained alive for much longer, he added.

"The next step is to see how the fetus develops after it has been taken out of the artificial womb," Kuwabara said.

Although his team has experimented for over 10 years, it was still early to hope for a quick end to miscarriages and premature baby fatalities through the use of an artificial womb," he added.

Appendix B

New Evidence of Prior Habitation of Human Life on Mars

Charles L. Spiegel

In 1976, the United States dispatched to Mars two robotic probes code-named "Project Viking." The mission consisted of four spacecraft for detailed investigations both on the surface and from orbit. In July of that year, the initial Orbiter (designated the "A" spacecraft) secured a high-resolution frame of a region called Cydonia as part of an intensive mapping sequence. Initial analysis of this frame by the Viking Imaging Team revealed an intriguing curiosity: a mile-wide mesa resembling a humanoid "head." Pictures of the region were subsequently released to members of the press by NASA's Jet Propulsion Laboratory. Official comment at the time (August 1976) dismissed any scientific significance, terming the resemblance merely a trick of lighting. Recent comments by members of the planetary community on the subject of the "Face" reveal no deeper interest or analy-

"NASA/JPL/Caltech"

The Face on Mars, located in the Cydonia Region

sis during the eight years since the image was secured by Viking.

In 1979, DiPietro and Molenaar rediscovered the image of the controversial feature in the National Space Science Data Center files. Subsequently, they embarked upon an independent program to locate additional *Viking* images and to perform image processing techniques on the original digital data that was used to construct those images in an effort to extract the maximum existing data. This would serve to better characterize the nature of this enigmatic feature. They were successful in their pursuit of both of these objectives.

In addition to the controversial Face, DiPietro and

Molenaar discovered a second unusual object on frame 70A13—a pyramid-like object. Located within a few kilometers of the previous anomaly, the object was characterized by four apparent sides, four flattened planes and peculiar "buttressing" at each apparent corner. The scale of this new feature was similar to the dimensions of the Face: 0.6 kilometers for the short sides and approximately 2.6 kilometers for the long sides. Data taken from the spacecraft data-block on 70A13 indicated that the western "pyramidal plane" was oriented parallel to the spin axis of the planet, i.e. north and south.

The point is that to even conceive of the unusual surface features identified by this team as being anything more than the products of exotic geological conditions requires the present inhospitable Martian environment to have been different at some time in the past, possibly in the recent past—a few million years ago. Curiously, in the existing geochemical data supplied by *Viking,* such a possibility is not excluded."

The puzzling presence of abundant oxides in the current Martian soil and their persistence in the face of conditions that are systematically destroying them is a "relic" of a recent, drastic change in the Martian environment.

The presence of the remnant polar caps—often composed totally of water, ice, and apparently solid carbon dioxide—remains one of the outstanding anomalies of Mars. Thus, in the youngest geologic features on the planet, we see evidence of abrupt changes in whatever mechanism led to the formation of the current polar caps.

Via cratering curves, the approximate epoch when this proposed change might have taken place can be calculated to have been, possibly, as recent as a few million years ago.

Other supportive evidence in favor of a "recent" change in paleo-climatic conditions for Mars comes from analysis of the circumpolar Great Sand Sea and from the presence of wind-eroded artifacts.

Finally, a long-accepted indicator of a potentially different Martian climate—albeit a very ancient climate—has been the runoff channels.

These arrays of apparently eroded river valleys have been cited often as evidence for a former "warm, wet epoch" in the history of Mars. But because of the observed preference of these features for the ancient cratered highlands, many investigators have almost automatically assumed that the creation of the channels (from erosive processes existent during a brief denser atmospheric epoch concurrent with the bombardment which produced the cratered highlands) confirms the presumption that the runoff channels were formed very early in the planet's history. Their existence as features restricted primarily to the oldest terrains on Mars has been seen as persuasive evidence that the warm, wet epoch was a concurrent by-product of this brief formation period—and disappeared soon after.

Two specific unmanned missions—the Soviet *Phobos* probe in 1988 and the U.S. *Mars Observer* spacecraft in 1990—have officially been scheduled. Representatives from the U.S. State Department and the Soviet Foreign Ministry, in November of 1986, concluded a framework agreement to undertake the cooperative scientific exploration of Mars and other planets. It specifically calls for the Soviet Union to share scientific data from its 1988 *Phobos* probe, which is to land on a Martian moon . . . and for the United States to share with Soviet scientists data returned from its *Magellan* probe to Venus, which is to be launched by the space shuttle in the spring of 1989.

Appendix C

Global Climatic Change on Mars

Today a frozen world,
Mars at one time may have had more temperate conditions, with flowing rivers, thawing seas, melting glaciers and, perhaps, abundant life.

Jeffrey S. Kargel and Robert G. Strom

To those of us who have spent a good part of our lives studying Mars, the newly discovered evidence that extraterrestrial microbes may have once lived in a rock cast off from that planet stirs feelings of awe. But the recent claim also evokes thoughts of Percival Lowell, a well-known American astronomer of the early 20th century, who turned his telescope toward Mars and saw a vast network of canals bordered by vegetation. His suggestion that Mars harbored such lushness had many people believing that the surface of the planet enjoyed conditions not so different from those on Earth. But in the 1960s three *Mariner* spacecraft flew by Mars and revealed the true harshness of its environment.

Observations from those unmanned probes indicated that Mars has an atmosphere that is thin, cold and dry. This tenuous shroud, composed almost entirely of carbon dioxide, provides less than one percent of the surface pressure found at sea level on Earth. The images radioed back during those first fleeting encounters three decades ago were fuzzy and few in number, but they were decidedly more accurate than Lowell's telescopic views. The *Mariner* cameras showed no canals, no water and no vegetation. They presented only a moonlike surface covered with craters. Sober scientists quickly dismissed any notion that the climate on Mars was sufficiently warm or wet to sustain life.

With its distant orbit—50 percent farther from the sun than Earth—and slim atmospheric blanket, Mars experiences frigid weather conditions. Surface temperatures typically average about -60 degrees Celsius (-76 degrees Fahrenheit) at the equator and can dip to -123 degrees C near the poles. Only the midday sun at tropical latitudes is warm enough to thaw ice on occasion, but any liquid water formed in this way would evaporate almost instantly because of the low atmospheric pressure.

Although the atmosphere holds a small amount of water, and water-ice clouds sometimes develop, most Martian weather involves blowing dust or carbon dioxide. Each winter, for example, a blizzard of frozen carbon dioxide rages over one pole, and a few meters of this dry-ice snow accumulate as previously frozen carbon dioxide evaporates from the opposite polar cap. Yet even on the summer pole, where the sun remains in the sky all day long, temperatures never warm enough to melt frozen water.

Despite the abundant evidence for cold, dry conditions, the impression of Mars as a perpetually freeze-dried world has been steadily giving way since the Mariner probes first reported their findings. Planetary scientists, who continue to examine the voluminous data from both the Mariner and the later Viking missions of the 1970s, now realize that Mars has had a complex climatic history—one that was perhaps punctuated with many relatively warm episodes. At certain times, huge volumes of water flowed freely across the surface of the planet. Before considering what this astonishing fact means for the possibility of life evolving on Mars or the strategy for the next round of Martian exploration (which is just now getting under way), it is instructive to review how this reversal in the way Mars is perceived came about.

Muddy Recollections

Scrutinizing the *Mariner* and *Viking* images obtained from orbit, planetary scientists soon noticed that most old Martian craters (unlike lunar ones) are eroded and that features resembling mudflows occur around almost every large, young crater on Mars. Such muddy "ejecta" probably represent the frozen remnants of a cataclysmic moment in the past when an asteroid or comet collided with the Martian surface, melting a patch of icy permafrost (where water-saturated ground had been frozen) and excavating a large hole that tapped a zone containing liquid water deep underground. By the late 1970s planetologists concluded that a considerable amount of underground ice and water has been present below the Martian surface throughout much of the history of the planet.

Yet not all Martian craters have these muddy flows surrounding them. Smaller craters appear more like their counterparts on the moon, with just streaks of dry ejecta scattered around them. Near the equator of Mars, only craters greater than about four kilometers in diameter display muddy ejecta, but closer to the Martian poles, craters as small as one kilometer across also have relic mudflows. This dependence on latitude arises because the ice-free, surficial layer varies in thickness. This layer extends deeper near the equator (to about 800 meters) than near the poles because the relative warmth of the Martian tropics purges much of the subsurface of frozen water. Hence, near the equator only the impact of bigger objects (that is, those that leave relatively large craters) will burrow down through the upper layer to heat the underlying icy permafrost and release a torrent of mud.

Researchers have since found other indications that a thick substratum of frozen ground exists on Mars. They have also identified evidence that ice once formed on the surface, where it appears to have created characteristic glacial landscapes. These features include bouldery ridges of sediment left by melting glaciers at their margins and meandering lines of sand and gravel deposited beneath glaciers by streams running under the ice (so-called eskers).

Many telltale landforms on Mars resemble frosty sites on Earth. For example, the pitted terrain on Mars corresponds to an earthly equivalent called thermokarst, which forms when the ice contained at shallow levels melts and the ground collapses. The apron-shaped lobes of rocky debris seen on the flanks of some Martian mountains might be rubble-covered glaciers. Or, more likely, they represent "rock glaciers," like the ones that form within the Alaska Range and in the Antarctic Dry Valleys on Earth. These distinctive sloping surfaces result after thousands of freeze-thaw cycles and cause the top meter or so of water-soaked ground to creep slowly downhill.

Glacial features and muddy ejecta around craters are not the only examples of water shaping the Martian surface. In some places, sinuous valleys one kilometer wide and many hundreds of kilometers long form large branching networks. Carl Sagan of Cornell University, Victor R. Baker of the University of Arizona and their colleagues suggested in the 1970s that such troughs were created by running water. Other Martian valleys have blunt starting points and short tributaries, characteristics that are typical of erosion by groundwater "sapping." That process, common on Earth, results from the seepage of water from underground springs, which causes the overlying rock and soil to wash away.

Images of Mars also reveal enormous outflow channels etched on the surface. Some of these structures are more than 200 kilometers wide and can stretch for 2,000 kilometers or more. These channels emanate from what is called chaotic terrain, regions of fractured, jumbled rocks that apparently collapsed when groundwater suddenly surged outward. The ensuing floods carved the vast channels, leaving streamlined islands more than 100 kilometers long and gouging cavernous potholes several hundred meters deep. Baker compared the Martian outflow channels to similar, albeit smaller, flood features found on Earth in parts of Oregon and Washington State. Those so-called channeled scablands of the Pacific Northwest formed after a glacier that had dammed a large lake broke open suddenly and caused a catastrophic flood.

The geometry of the Martian outflow channels indicates that water could have flowed along the surface as rapidly as 75 meters per second (170 miles per hour). Michael H. Carr of the U.S. Geological Survey estimates that the vast quantity of water necessary to create these many enormous channels would have been enough to fill a global Martian ocean that was 500 meters deep, although not all this liquid flowed at one time. One source for that great quantity of water may have been a deep lake in Valles Marineris, a region on Mars partly covered with sedimentary layers that appear to be ancient lake deposits. Water could also have gushed from a large reservoir under ice-impregnated permafrost that had been warmed by heat from the interior of the planet.

Why should such an underground accumulation of water suddenly inundate the surface? Scientists are unsure of the exact cause, but this groundwater might have started to flow after the icy permafrost capping it thinned and weakened, perhaps because of a sudden climate warming, volcanism or tectonic uplift. Perhaps a large meteor impact or quake triggered the cataclysmic dousing. Once water broke through to the surface, carbon dioxide from saturated groundwater—a Martian seltzer of sorts—may have erupted in tremendous geysers, further undermining the stability of the saturated underground layers. The result was to produce chaotic terrain and to unleash floods and mudflows of a magnitude that has rarely, if ever, been matched by any earthly deluge.

An Ocean Away

Some highland areas on Mars contain extensive systems of valleys that drained into sediment-floored depressions. These lowlands were at one time full of water. The largest of these Martian lakes filled two gigantic impact basins called Hellas and Argyre.

But these lakes may not have been the largest bodies of water on the planet. Research groups led by David H. Scott and Kenneth L. Tanaka of the U.S. Geological Survey and by Jeffrey M. Moore of the National Aeronautics and Space Administration Ames Research Center independently concluded that repeated floods from the outflow channels emptied to the north and formed a succession of transient lakes and seas. We have interpreted many features bordering these ancient basins as marking where glaciers once emptied into these deep bodies of water. Tanaka and Moore believe that thick layers of sediment deposited in these seas now stretch across much of the extensive northern plains. According to several estimates, one of the larger of the northern seas on Mars could have displaced the combined volume of the Gulf of Mexico and the Mediterranean Sea.

Yet even that great body of water may not have been the supreme example: there may have been a Mars ocean. As early as 1973 the late Henry Faul of the University of Pennsylvania raised this intriguing possibility in a paper he romantically entitled "The Cliff of Nix Olympica." Understandably, given the paucity of observations then available, the paper was never accepted for publication. But during the past decade, other researchers, working with information acquired during the Viking missions, have revived Faul's idea.

For instance, in 1989 Timothy J. Parker and his colleagues at the Jet Propulsion Laboratory in Pasadena, California, again proposed a northern ocean (arguing that many features in the northern plains looked as if they had resulted from coastal erosion). To enhance prospects for publication, however, they deliberately obscured the provocative thrust of their work with the mundane title "Transitional Morphology in the West Deuteronilus Mensae Region of Mars: Implications for Modification of the Lowland/Upland Boundary." In a subsequent paper, these researchers ventured a more direct title to convey their ideas: "Coastal Geomorphology of the Martian Northern Plains." Motivated in part by such work, Baker and several colleagues (including us) named this hypothetical northern ocean Oceanus Borealis. We calculated that it was possibly four times as large as the Arctic Ocean on Earth, and we proposed a scenario for the actions of the water cycle on Mars that could have accounted for it.

Whereas most planetary scientists now agree that large bodies of water formed repeatedly in the northern plains on Mars, many do not accept that there was ever a true ocean there. Some envision that only a vast, muddy slurry, or mud ocean, existed. In any case, it is clear that huge amounts of water once flowed over the surface of Mars. Yet the fate of that water remains unknown. Some of it may have percolated into the subsurface and frozen in permafrost. Some may have frozen in place and might now stretch across much of the floor of the northern plains, hidden by a mantle of dust and sand. Some water may simply have evaporated, to be later lost to space or deposited as snow at the poles.

Trust the Old Salts

Although images of the landforms left by ancient glaciers, river valleys, lakes and seas are strong testament that Mars was once rich in water, evidence comes from other sources as well. Earth-based spectroscopic measurements of Mars reveal the presence of clay minerals. Even more directly, the two landers that set down on the surface during the Viking program analyzed Martian soil and found that it probably contains 10 to 20 percent salts. Martian rocks, like those on Earth, react to form salt and clay minerals when exposed to water. But such chemical weathering probably cannot occur under the cold and dry conditions that now reign on Mars.

Some scientists have also studied Martian rocks found here on Earth. These rare samples of the Martian surface were blasted into space by the impact of an asteroid or comet and later fell to Earth as meteorites. Allan H. Treiman of the Lunar and Planetary Institute in Houston and James L. Gooding of the NASA Johnson Space Center have shown in the past several years that minerals in some of these so-called SNC meteorites were chemically altered by cool, salty water, whereas others were affected by warmer hydrothermal solutions. Their conclusions imply that Mars once had a relatively warm, wet climate and may have had hot springs. Just perhaps, conditions were right for life.

That possibility inspired David S. McKay of the NASA Johnson Space Center and his colleagues to examine an SNC meteorite for signs of ancient Martian life. Although their conclusion that fossil microbes are present is open to debate (and a vigorous one is indeed going on), the composition of the rock they studied—with fractures filled by minerals that probably precipitated from an aqueous solution—indicated that conditions on Mars a few billion years ago would have been compatible with the existence of life.

In agreement with this assessment, many atmospheric

physicists had already concluded that Mars has lost immense quantities of water vapor to space over time. Their theoretical calculations are in good accord with measurements made by various Soviet space probes that showed oxygen and hydrogen atoms (derived from breakdown of atmospheric water exposed to sunlight) streaming away from Mars. The continuous loss of these elements implies that Mars must once have had all the water needed to fill an Oceanus Borealis.

But water was not the only substance lost. Recently David M. Kass and Yuk L. Yung of the California Institute of Technology examined the evolution of carbon dioxide—a potent greenhouse gas—in the atmosphere of Mars. They found that over time an enormous quantity of carbon dioxide has escaped to space. That amount of gaseous carbon dioxide would have constituted a thick Martian atmosphere with three times the pressure found at the surface of Earth. The greenhouse effect from that gas would have been sufficient to warm most of the surface of Mars above the freezing point of water. Thus, from this perspective, too, it seems quite plausible that the climate on Mars once was much warmer and wetter than it is today.

Yet many questions remain about how water might have arranged itself on the surface of Mars: Was there actually an ocean? Did water shift rapidly between different reservoirs? When and for how long was Mars wet? Although the absolute timing of these events remains unknown, most researchers believe that water sculpted the surface of Mars at many intervals throughout the history of the planet. The constant loss of water and carbon dioxide from the atmosphere suggests that early epochs on Mars (that is, billions of years ago) may have been especially warm and wet. But some balmy periods may also have been relatively recent: Timothy D. Swindle of the University of Arizona and his colleagues studied minerals in an SNC meteorite created by aqueous alteration and determined that they formed 300 million years ago—a long time by human standards but only a few percent of the age of the 4.6-billion-year-old solar system. Their result was, however, accompanied by a considerable degree of uncertainty.

The duration of the wet periods on Mars is also difficult to gauge exactly. If the eroded Martian landscapes formed under conditions typical of terrestrial glacial environments, more than a few thousand but less than about a million years of warm, wet climate were required. Had these conditions endured substantially longer, erosion would have presumably erased all but traces of a few impact craters, just as it does on Earth.

This limitation does not apply to the earliest history of the planet, billions of years ago, before the craters now visible had formed. A young Mars may well have had vigorous erosion smoothing its face. But eventually, as the planet slipped toward middle age, its visage became cold, dry and pockmarked. Only scattered intervals of warmth have since rejuvenated the surface of the planet in certain regions. Yet the mechanism that causes Mars to switch between mild and frigid regimes remains largely mysterious. Scientists can now venture only crude explanations for how these climate changes might have occurred.

Turning on the Heat

One hypothesis involves shifts in obliquity, the tilt of the spin axis from its ideal position, perpendicular to the orbital plane. Mars, like Earth, is now canted by about 24 degrees, and that tilt changes regularly over time. Jihad Touma and Jack L. Wisdom of the Massachusetts Institute of Technology discovered in 1993 that, for Mars, the tilt can also change abruptly. Excursions of the tilt axis through a range of as much as 60 degrees may recur sporadically every 10 million years or so. In addition, the orientation of the tilt axis and the shape of the orbit that Mars follows both change cyclically with time.

These celestial machinations, particularly the tendency of the spin axis to tilt far over, can cause seasonal temperature extremes. Even with a thin atmosphere such as the one that exists today, summer temperatures at middle and high Martian latitudes during periods with large obliquity could have climbed above freezing for weeks on end, and Martian winters would have been even harsher than they are currently.

But with sufficient summer warming of one pole, the atmosphere may have changed drastically. Releases of gas from the warmed polar cap, from seltzer groundwater or from carbon dioxide-rich permafrost may have thickened the atmosphere sufficiently to create a temporary greenhouse climate. Water could then have existed on the surface. Aqueous chemical reactions during such warm periods would in turn form salts and carbonate rocks. That process would slowly draw carbon dioxide from the atmosphere, thereby reducing the greenhouse effect. A return to moderate levels of obliquity might further cool the planet and precipitate dry-ice snow, thinning the atmosphere even more and returning Mars to its normal, frigid state.

This theory of climatic change needs to be tested, but new observations and fresh insights will undoubtedly come from a decade-long series of unpiloted spacecraft that will next visit Mars. The expeditions begin this month with the launch of American and Russian probes. This program of exploration had been slated to conclude in 2005 with the return of Martian rocks. But the discovery of what may be fossil microbes in an SNC meteorite has sparked thoughts of obtaining Martian samples sooner so that scientists can better evaluate whether microorganisms existed on Mars several billion years ago—or even more recently.

The American spacecraft soon to be under way include *Mars Pathfinder* and *Mars Global Surveyor. Pathfinder* will land on a bouldery plain of an outflow channel that once fed an ancient sea. Although not equipped to test directly for signs of life, this lander will release a small roving vehicle to explore the local environs. *Surveyor* will take pictures from orbit that can resolve features that are

"NASA/JPL/Caltech"

Lander image of rover near The Dice (three small rocks behind the rover) and Yogi on sol 22

just a few meters across. Measurements from this orbiter will also allow scientists to make detailed topographic maps and to search for icy deposits as well as new evidence of ancient glaciers, lakes and rivers. Information gathered by these next missions should give scientists a clearer picture of what Mars looked like during its last episode of warmer climate, perhaps 300 million years in the past.

By 300 million years ago on Earth, amphibians evolved from fish had crawled out of the sea and inhabited swampy coastlines. Might other complex creatures have flourished simultaneously along Martian shores? The basic conditions for life may have existed for a million years late in Martian history—perhaps much longer during an earlier period. Were these intervals conducive for organisms to evolve into forms that could survive the dramatic changes in climate? Could Martian organisms still survive today in underground hot springs? The next decade of concentrated exploration may provide the definitive answers, which, if positive, would mark an intellectual leap as great as any in human history.

Further Reading

"Ancient Oceans, Ice Sheets and the Hydrological Cycle on Mars." V. R. Baker, R. G. Strom, V. C. Gulick, J. S. Kargel, G. Komatsu and V. S. Kale in *Nature*, Vol. 352, pages 589-594; August 15, 1991.

"Ancient Glaciation on Mars." J. S. Kargel and R. G. Strom in *Geology*, Vol. 20, No. 1, pages 3-7; January 1992.

"The Ice Ages of Mars." J. S. Kargel and R. G. Strom in *Astronomy*, Vol. 20, No. 12, pages 40-45; December 1992.

"Coastal Geomorphology of the Martian Northern Plains." T. J. Parker, D. S. Gorsline, R. S. Saunders, D. C. Pieri and D. M. Schneeberger in *Journal of Geophysical Research E (Planets)*, Vol. 98, No. 6, pages 11061-11078; June 25, 1993.

Appendix D

An Unearthly Landscape, Yet Somewhat Familiar

K. C. Cole

The allure of life on the planet next door has tugged at Earth's imagination for centuries. And yet, admits Donna Shirley, manager of NASA's Mars program, "we know very little about Mars."

Earthlike in some ways but otherworldly in others, Mars is small, with less than half the gravitational attraction of Earth. It has no atmospheric blanket to trap the heat, so the surface gets icy cold. It has no ozone umbrella to block the ultraviolet solar energy.

The Viking spacecraft, which visited Mars in 1976, discovered no signs of life but did find a highly reactive soil chemistry that chews up organic molecules before they have a chance to stick together. The planet's surface is more sterile than many Earth laboratories. If there is life anywhere, it would have to be hiding deep beneath the surface. Every trace of carbon appears to be locked in the atmosphere.

In other ways, however, the pink landscape looks a lot like home, with rolling sand dunes embellished with fine ripples, boulder-strewn fields reminiscent of the Wild West, and 50,000-foot high volcanic peaks. The soil is the same heft as good farming soil on Earth.

The planet has two polar caps of frozen carbon dioxide and water, and possibly Mars-quakes similar to earthquakes, though not as frequent. The ground is scarred by parallel fractures that look a lot like Earth's fault lines.

One major difference is that Mars appears to be bone-dry—despite clear evidence that water once flowed freely on its surface. Finding out what happened to the water will be the highest priority in this and future missions, because only water has the remarkable chemical and physical properties that make life as we know it possible.

Appendix E

The Evidence for Liquid Water on Mars

Donald Goldsmith

Because Mars was born small, with only half the Earth's diameter and one-tenth of Earth's mass, it fell short of Earth's ability to capture and retain volatile molecules such as nitrogen, H_2O, and carbon dioxide. These are the compounds that gave Earth and Mars their atmospheres. Mars was born comparatively poor in volatiles, and grew still poorer as it aged. Four billion years ago, Mars had liquid water on its surface, but as the planet grew older, its water gradually disappeared.

How do we know that Mars had liquid water in bygone eras? The proof came two and a half decades ago, when the *Mariner* spacecraft photographed the planet's entire surface. These photographs, confirmed in more detail by the *Viking* orbiters, show sinuous channels, which geologists feel sure must have been carved by liquid in motion, as well as dried-up lake beds and at least one example of a sudden flow of water on a large scale. The only liquid that is likely ever to have existed on Mars is water, whose molecules are made from two of the most abundant elements in the solar system. Spectroscopic measurements of the sunlight reflected from the Martian polar caps show that they consist mainly of frozen carbon dioxide, but a fraction of a percent of their total volume consists of water, ice, and a tiny amount of water vapor which exists in the planet's thin atmosphere.

Mars almost certainly never had planet-wide oceans, as Earth did and to a large extent still does. Four billion years ago, however, Mars had liquid water coursing across its surface and possibly raining from its skies. The atmosphere must have been much thicker then, capable of providing both the larger greenhouse effect and the higher surface pressure needed to allow liquid water to exist on Mars.

This atmosphere presumably arose from volcanic activity. Today the Martian volcanoes are dormant or extinct, though their enormous bulk makes them the tallest mountains in the solar system, mute evidence that Mars' surface once cracked and moaned with volcanic activity that released hydrogen, carbon monoxide, carbon dioxide, and nitrogen gases trapped in the planet's subsurface layers. Even greater amounts of these gases should have arrived from space, trapped in icy planetesimals and comets, during the era of intense bombardment that ended 3.8 billion years ago.

As a result of this bombardment and the release of gases by volcanoes, billions of years ago, Mars should have had a thick atmosphere made of two gases: carbon dioxide, which forms the bulk of the present atmosphere of Mars, and water vapor. If these two substances had an abundance sufficient to increase the current atmospheric pressure by a factor of ten, water could have existed as a liquid within the temperature range from 32 to 70 degrees Fahrenheit (0 to 21 degrees Celsius). If instead the Martian atmosphere contained one hundred times the amount of carbon dioxide that it does now, the atmospheric pressure would have risen to 60 percent of the surface pressure on Earth, extending the temperature range for liquid water to span the interval from 32 to 185 degrees Fahrenheit (0 to 85 degrees Celsius). Planetary experts think that the primitive Martian atmosphere was still thicker, perhaps as thick as Earth's is now. If the additional atmospheric pressure arose not only by carbon dioxide but also by water vapor in the atmosphere, so much the better: Water's ability to remain liquid depends only on the total atmospheric pressure, and a larger amount of water vapor makes a cycle of evaporation and rainfall more likely.

Mars: That Other Eden, Demi-Paradise

Thus primitive Mars, the planet as it was more than four billion years ago, when it was less than half a billion years old, had abundant liquid water and an atmosphere dozens or hundreds of times thicker than it is now. Although scientists do not concur as to whether complete cycles of rainfall, evaporation, and further rainfall occurred on primitive Mars, they do agree that liquid water carved the channels, still visible four million years later, and once filled some of the craters whose interiors show ancient shorelines. Primitive Mars was not only wetter than Mars today, but warmer as well. The greater amounts of atmospheric carbon dioxide and water vapor would have increased the "greenhouse effect," the ability of the atmosphere to trap infrared radiation from the surface, and thus to keep the planet warm. Even today, points on the Martian equator reach noontime temperatures of 60° or 70° Fahrenheit (15° to 21° Celsius), though they cool to more than 100 degrees below zero by midnight, so poorly does the thin atmosphere retain heat. With an atmosphere ten times thicker than the current one, Mars could have had slightly higher daytime temperatures and much warmer temperatures at night.

In Shakespeare's *King Richard II*, the monarch's uncle, John of Gaunt, speaks of "this earth of majesty, this seat of Mars/This other Eden, demi-paradise." Even though Gaunt's "seat of Mars" was in fact a reference to England, the characterization of Mars as "this other Eden, demi-paradise" serves well to describe the scene on the Martian surface, four billion years ago, when rain fell, water flowed, volcanoes belched, and the conditions for life were all in place.

Where did the Water Go?

What happened to this demi-paradise of primitive Mars? Somehow, the atmosphere steadily grew thinner. Some of the carbon dioxide may have escaped into space, and a small fraction of the atmospheric CO_2 froze into the Martian polar caps. Some of the water vapor likewise escaped, while a tiny amount joined the dry ice of the polar caps; and the bulk of the water, quite possibly, remains widely distributed over the planet, not on but under the Martian surface.

Wherever the carbon dioxide and water vapor may have gone, the fact they disappeared almost totally from Mars' atmosphere has had a sad effect on the Martian environment. As the atmospheric pressure decreased, so too did the amount of the atmospheric greenhouse effect, because the atmosphere contained fewer molecules that prevented heat from escaping from the planet's surface. As the greenhouse effect decreased, Mars lost not only its opportunity to have liquid water but also its ability to remain relatively warm at night. This increased the rate at which water disappeared from the Martian surface, driving it into either the permafrost or the polar caps.

The net effect was to turn Mars from a primitive planet with liquid water to a totally dry world, devoid of any chance of liquid. This change probably took no longer than a few hundred years, at a time somewhere between three and four billion years ago. During that era, it was good-bye to a Mars fit for life, and hello to a planet hostile to life as we know it.

What we know and have concluded about the origin of life on Earth strongly suggests that primitive Mars may well have been nearly as ready for life as primitive Earth was. If life did begin on primitive Mars at a time close to four billion years ago, the burning question that confronts us today is: Did life totally disappear on the Red Planet or did life find a few places of refuge, where even today it might persist, awaiting our discovery?

If life requires liquid water, as all life on Earth does, then we cannot expect to find life on the Martian surface. But this does not eliminate all the possibilities for Martian life. We can envision forms of life that have found niches beneath the polar caps, or even better, in subterranean caverns heated by volcanic activity. These "underground oases" are entirely imaginary for now. But they rank high on the list of items that scientists would like to investigate in any return to Mars. Temperature-sensing satellites might be able to locate underground regions significantly warmer than their surroundings, and robot explorers might be able to drill into what might prove underground reservoirs where forms of life have survived nearly four billion years despite the absence of liquid water on the Martian surface.

Enough of these "mights"! Think of the excitement and the frustration of scientists who have dreamed of finding life on Mars, have lived through the twenty years that followed the negative results from the *Viking* missions to Mars and have now learned from ALH 84001 that life may have existed on Mars three to four billion years ago.

Only a few dozen million miles separate them from the chance to determine whether any life exists on Mars now, and if it does, to analyze those forms of life to discover their differences from, and resemblances to, Earth life.

Extraterrestrial Life Throughout the Universe?

One result of the discovery of new planets and possible ancient life on Mars seems particularly striking: The likelihood of extraterrestrial life has increased not in reality, which remains as it was before the recent discoveries, but in our estimates of the probabilities. Before late 1995, astronomers felt sure that sun-like stars must have planets, as nothing about our sun distinguishes it from its many thousands of near-twins in the Milky Way. On the other hand, no one had *detected* any of these putative planets, a fact that could be explained by the technical difficulties involved, but that was nonetheless troubling. Finding the new planets has lifted a mental burden, leaving astronomers one step closer to obtaining proof that life, and perhaps intelligent life at that, is widespread in the universe.

The news from Mars likewise cheers scientists who believe that Earth-like conditions elsewhere in the universe should have produced, in most if not all cases, other forms of life that inhabit worlds other than ours. Although ALH 84001 does not contain definitive proof of life on Mars, this ancient Martian rock does seem to make more likely the possibility that Mars once teemed with life. Should this turn out to be true, theories that imply that life has arisen on planet after planet will receive validation. If two neighboring planets in the same system possess histories that include life, the argument that life arises only rarely, and under highly specific conditions, vanishes into thin air. One might attempt to resurrect it by invoking panspermia, but this would only prove that even though the actual origin of life may be a rare event, life spreads relatively easily from place to place in the cosmos. Most scientists and most of the general public would consider other forms of life quite fascinating even if they evolved from cosmic seeding. In particular, we would be highly interested in whether extraterrestrial forms of life, no matter how they originated, have evolved to our stage of development, or even beyond.

On this point we remain almost as ignorant as we were before the discoveries of extrasolar planets and possible ancient life on Mars. If we increase our estimates of the number of worlds that are favorable to life, and of the number of those worlds on which life has appeared, this might seem to require an increase in our estimate of the number of places that might have given rise to a society capable of, and interested in, communication across interstellar distances.

On the other hand, we have no good evidence that any such societies exist. The evidence adduced in favor of extraterrestrial visitors to Earth consists either of individual memories and observations, or of physical evidence similar to the Face on Mars, easily seen to be far from anything that would convince a reasonable jury. Although eyewitness testimony is admissible in most courtroom proceedings, one can argue that when dealing with strange events, such accounts have proven so unreliable that they cannot be trusted. The weirder the alleged occurrence, the less one can rely on eyewitnesses, a fact well known to psychologists, though anathema to the human psyche. In the movie *Duck Soup,* Chico Marx asks the rich lady, "Who are you going to believe—me or your own eyes?" And he is right to do so, at least in this case, for he is telling the truth: it was not Chico, but Harpo, whom the lady saw. From a scientific viewpoint, you can believe neither me nor your own eyes. Until we have hard physical evidence of extraterrestrial visits, something a good deal more convincing than, for example, crop circles that hoaxers can easily make, we should reject amazing hypotheses in favor of more mundane ones.

The views above are, of course, entirely my own (unless they could lead to trouble with believers in a conspiracy to suppress UFO information, in which case I have no idea whose they are). From a scientific viewpoint, estimating the existence of highly advanced civilizations elsewhere in the cosmos rests about where extrasolar planets did some years ago: On the basis of the large numbers of possible sites for life, scientists think advanced civilizations *should* exist, but have no convincing evidence that they do. The *absence* of documented extraterrestrial visitors to Earth, along with the failure to detect any radio signals from extraterrestrial civilizations, argues in a modest way against the existence of other civilizations, at least in our corner of the Milky Way galaxy.

However, we should remember that the same fact that argues in favor of a large number of sites for life—the enormous volumes of space in the universe, which contains huge numbers of stars with, presumably their own planets as well—also argues against interstellar travel. Interstellar distances dwarf those in our solar system: The closest stars to the sun are a million times farther from us than Mars at its closest. If we make a model of the Milky Way in which stars have the sizes of light bulbs, each of these lights would have a large city to itself, with one bulb each in New York, Boston, London, Munich, Athens, and Delhi. Although we cannot predict how rich other civilizations may be in energy terms, physical realities suggest that they would start their attempts to contact any neighbors not with incredibly long journeys in person, but rather with radio signals. In that case, humanity owes it to itself to increase the relatively modest efforts that are now under way to listen, so far without success, for possible signals from intelligent forms of life around other stars.

From the discovery of a piece of rock on ice to the quest for other civilizations may seem a large step, but when the rock comes from another planet, was formed more than four billion years ago, and may contain material formed by ancient extraterrestrial life, the leap seems less gigantic. Only time can tell whether we shall look back upon 1996 as the year in which the first extraterrestrial life was discovered, billions of years after it lived, millions of years after its remnants left its planet, and thousands of years after those remnants fell to Earth.

Whatever the resolution of its enigmatic contents, ALH 84001 has, I hope, furnished my readers with a glorious ride through the cosmic story of life, so far as we know it today. With any luck, the time will soon come when we not only comprehend what the meteorite from Mars has to tell us, but also understand the origin and development of life in the solar system. This history awaits our discovery—if we can create the instruments that will "prospect" for life, send them to explore places where past or present life seems reasonable, and use them to analyze their discoveries with an organized skepticism that will allow us to discern the truth about life in the universe. Thanks for traveling with me on part of the journey; the rest is spread out before you.

Appendix F

From Deluge to Desert

Arid, airless, Mars once was awash, Pathfinder Evidence Shows

Mark Carrreau

PASADENA—Mars was not so hard on the eyes 3.5 billion to 4.6 billion years ago.

Today, arid cold and nearly airless, the planet is rusting away like an old automobile. But sometime between 1 billion and 3 billion years ago, it flooded with a biblical fury that would make this spring's floods in North Dakota appear no worse than a dribble from a garden hose.

The rich tapestry of images flowing to Earth yesterday from NASA's Mars Pathfinder mission paint a grim scene.

"It's very hard to visualize, but this flood was huge," said Pathfinder geologist Mike Malin, head of Malin Space Science Systems in La Jolla.

"But we don't know where the water went."

Scientists are slowly teasing out inferences about the red planet and its complex past from the pictures and information streaming to Earth from Pathfinder and its roving companion, Sojourner.

Geologists have known since the Viking missions 21 years ago that giant floods once swept the now-dry planet. But the Pathfinder pictures provide the most powerful evidence yet.

Many of the conclusions come from studying the varied hues of orange, battleship gray, and blue that dominate the pictures. The colors offer insight into mineral composition of the soil and rocks.

Pathfinder bounced to a landing on Friday in Ares Vallis in the planet's northern equatorial belt. It has been taking pictures and relaying messages to the Sojourner robot as it crawls the planet's surface, approaching rocks and sending back information.

The Ares Vallis site was chosen because pictures from previous missions showed that it was probably a flood plain scattered with rocks washed down from nearby ancient highlands.

Scientists think the highlands are 4.5 billion to 3.6 billion years old, which is the same period in which life emerged on Earth. The early evidence indicates that the most recent flood at the landing site occurred 1 billion to 3 billion years ago.

Yesterday, Sojourner was rolled to a rock named "Yogi," because it resembles a bear.

During daylight periods that last a little longer than

those on Earth, the sun blazes brightly, illuminating the terrain as if it were a perfect movie set.

Of special note is the red pigment in the fine soil and dust that cling to rocks and boulders across the desolate landscape surrounding the probe.

"The conclusion we are coming to is these bright red dust deposits are very enriched in oxidized iron," said Cornell University chemist, James Bell. "The surface of Mars is rusting. We don't know when it started rusting. We don't know how fast it's rusting."

To rust, though, the planet must have a source of oxygen. The primary constituent of the thin Martian atmosphere now is carbon dioxide. It could be a source of oxygen, as could water, which is comprised of hydrogen and oxygen.

According to scientists, there are a couple of possible explanations for the pervasive redness on Mars.

In one, the gray-blue rocks were formed from volcanic material and washed to Ares Vallis by flood waters or blasted to the surface by meteorite strikes. Their rusting occurred slowly over a vast passage of time.

A second possibility is that Mars once had a much different environment that was kinder to the rocks. But some unknown environmental change triggered the rusting.

"The answers to these questions have very important implications for the climate history of Mars," Bell said.

Malin, who has spent years studying the geology of catastrophic floods and volcanic eruptions on Earth, said,

"I'm in hog heaven, looking at these pictures of Mars."

Among the first things he noticed, which had not been apparent in earlier pictures, were patterns of grooves in the landscape south of the Pathfinder lander. They were presumably caused by the flood's scouring and redeposition of rocks and soil.

The ridges are about 10 feet high and about 70 feet apart. Malin said these could provide clues to the depth and velocity of the flood waters across the plain.

Another question regarding the flooding remains unanswered. Were the floods primarily rampaging water carrying some soil and rocks, like an overflowing Mississippi River? Or were they more like mud flows, about half water and half soil and rock? The latter, Malin said, could have transported greater amounts of huge boulders.

Examining pictures of dry channels near the Pathfinder landing site, scientists have measured canals 15 feet deep and 60 feet apart. Images of the soil indicate that after the massive flood, some water remained in puddles and eventually evaporated, leaving silt.

Sojourner's principle scientific instrument is an alpha proton X-ray spectrometer, designed to bombard a rock with protons and record changes in them as they return to the instrument. In this way, the spectrometer should be able to determine any elements in a rock's minerals, except for hydrogen and helium. The results will be the first determination of the composition of Martian rocks.

Pathfinder provided a weather report for the next

Mars day (beginning yesterday afternoon California time): afternoon highs of 10 degrees Fahrenheit with variable light winds, with a low temperature overnight of minus 105 degrees.

Photos taken by the Pathfinder lander's camera, which popped up to its full height Saturday night, showed for the first time how the surrounding landscape would look to an adult standing on the Martian surface.

The lander's camera has taken hundreds of photographs, many of them in color and 3-D.

Appendix G

Mars Pathfinder Results From JPL

Diane Ainsworth

Based on the first direct measurements ever obtained of Martian rocks and terrain, scientists on NASA's Mars Pathfinder mission report that the red planet may have once been much more like Earth, with liquid water streaming through channels and nourishing a much thicker atmosphere.

Among the more significant discoveries of the Mars Pathfinder mission was the identification of possible conglomerate rocks, which suggests the presence of running water to smooth and round the pebbles and cobbles, and deposit them in a sand or clay matrix, says Dr. Matthew Golombek, Mars Pathfinder project scientist at NASA's Jet Propulsion Laboratory, Pasadena, California. This scenario supports the theory that Mars was once warmer and wetter.

"If you consider all of the evidence we have at Ares Vallis—the rounded pebbles and cobbles and the possible conglomerate, the abundant sand- and dust-sized particles and models for their origins, in addition to the high silica rocks," Golombek says, "it suggests a water-rich planet that may have been more Earth-like than previously recognized, with a warmer and wetter past in which liquid water was stable and the atmosphere was thicker."

A panoramic view of Pathfinder's Ares Vallis landing site reveals traces of this warmer, wetter past, showing a flood plain covered with a variety of rock types, boulders, rounded and semi-rounded cobbles and pebbles. These rocks and pebbles are thought to have been swept down and deposited by floods which occurred early in Mars' evolution in the Ares and Tiu regions near the Pathfinder landing site.

This image taken by the Imager for Mars Pathfinder looks to the southwest, a cluster of large angular rocks tiled in a downstream direction from the floods. The south peak, of two hills known as Twin Peaks, can be seen on the horizon, about 1 kilometer (6/10ths of a mile) from the lander. The rocky surface is comprised of materials washed down from the highlands and deposited in this ancient outflow channel by a catastrophic flood.

"Before the Pathfinder mission, knowledge of the kinds of rocks present on Mars was based mostly on the Martian meteorites found on Earth, which are all igneous rocks rich in magnesium and iron and relatively low in silica," Golombek and a team of Pathfinder scientists report in a paper entitled, "Overview of the Mars Pathfinder Mission and Assessment of Landing Site Predictions."

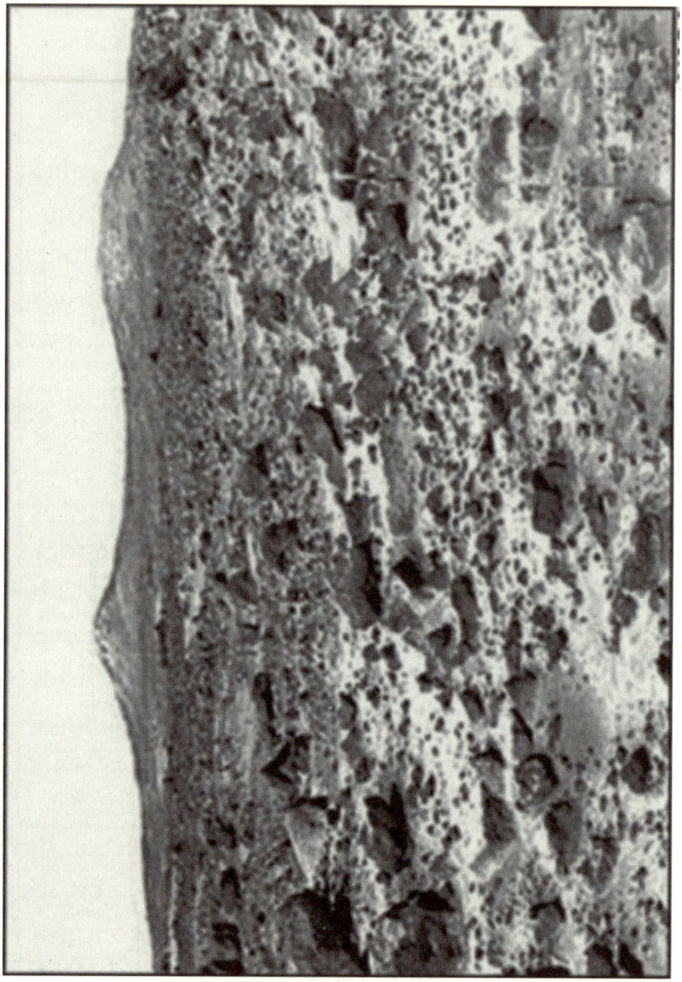

'JPL/NASA/Caltech'

The Twin Peaks as seen from the Ares Vallis site.
Picture returned by the *Mars Pathfinder* and the *Sojourner* rover.

The scientists report that chemical analyses of more than 16 rocks and studies of different regions of soil—along with spectral imaging of rock colors, textures and structures—have confirmed that these rocks have compositions distinct from those of the Martian meteorites found on Earth.

"The rocks that were analyzed by the rover's alpha proton X-ray spectrometer were basaltic or volcanic rocks, with granite-like origins, known as andesitic rocks," Golombek reports. "The high silica or quartz content of some rocks suggests that they were formed as the crust of Mars was being recycled, or cooled and heated up, by the underlying mantle. Analyses of rocks with lower silica content appear to be rich in sulfur, implying that they are covered with dust or weathered. Rover images show that some rocks appear to have small air sacks or cavities, which would indicate that they may be volcanic. In addition, the soils are chemically distinct from the rocks measured at the landing site."

The remarkably successful Mars Pathfinder spacecraft, part of NASA's Discovery program of fast-track, low-cost missions with highly focused science objectives, was the first spacecraft to explore Mars in more than 20 years. In all, during its three months of operations, the mission returned about 2.6 gigabits of data, which included more than 16,000 images of the Martian landscape from the lander camera, 550 images from the rover and about 8.5 million temperature, pressure and wind measurements.

The rover traveled a total of about 100 meters (328 feet) in 230 commanded maneuvers, performed more than 16 chemical analyses of rocks and soil, carried out soil mechanics and technology experiments, and explored about 250 square meters (820 square feet) of the Martian surface. The flight team lost communication with the lander on September 27, after 83 days of daily commanding and data return. In all, the lander operated nearly three times its design lifetime of 30 days, and the small, 10.5 kilogram (23 -pound) rover operated 12 times its design lifetime of seven days.

Now known as the Sagan Memorial Station, the Mars Pathfinder mission was designed primarily to demonstrate a low-cost way of delivering a set of science instruments and a free-ranging rover to the surface of the red planet. Landers and rovers of the future will share the heritage of spacecraft designs and technologies first tested in this "pathfinding" mission.

Golombek points out that the rocky surface and rock types found in Ares Vallis match the characteristics of a flood plain on Earth, created when a catastrophic flood washed rocks and surface materials from another region into the basin. Ares Vallis was formed in the same way that the 40-kilometer-long (25-mile) Ephrata Fan of the Channeled Scabland in Washington State was formed, says Golombek, adding that the Ephrata Fan was deposited when channels of water flowing down the Grand Coules filled the Quiney Basin.

Additional data from the Pathfinder landing site revealed that magnetic dust in the Martian atmosphere has been gradually blanketing most of the magnetic targets on the lander over time. "The dust is bright red, with magnetic properties that are similar to that of composite particles," Golombek states. "A small amount of the mineral maghemite has been deposited almost like a stain or cement. These results could be interpreted to mean that the iron was dissolved out of crustal materials in water, suggesting an active hydrologic cycle on Mars. The maghemite stain could be a freeze-dried precipitate."

Another team of scientists used daily radio Doppler tracking and less frequent two-way radio ranging techniques during communications sessions with the spacecraft to pinpoint the location of the Pathfinder lander in inertial space and the direction of Mars' rotational axis.

In his published paper, Dr. William Folkner, an interdisciplinary scientist at JPL, and co-authors present estimates of the Martian polar moment of inertia, which show that Mars has a dense core surrounded by a lighter mantle. The results imply that the radius of Mars' core is larger than about 1,300 kilometers (807 miles) and less than about 2,400 kilometers (1,490 miles). Mars' core and mantle are probably warmer than Earth's at comparable depths. Eventually, scientists may be able to determine whether Mars' core is presently molten or fluid.

"Variations in Mars' rotation around its own spin axis are thought to be dominated by mass exchange between the polar caps and the atmosphere," Folkner reports. "During winter, part of the atmosphere condenses at the poles. If the southern cap increased symmetrically as the northern cap decreased, then there would not be any change in moment of inertia or rotation rate. However, because of Mars' orbital eccentricity, difference in elevation and difference in albedo, the polar caps are not formed symmetrically.

"The unbalanced waxing and waning of the Martian polar ice caps results in seasonal changes in air pressure at the Pathfinder and Viking landing sites," he says. "These changes in air pressure are correlated with changes in Mars' rotation rate, and have been observed in our radio tracking measurements."

The season and time of arrival of Mars Pathfinder in the late northern summer resulted in some variations in the temperature of the upper atmosphere compared to Viking data, reports Dr. Tim Schofield, JPL team leader of the atmospheric structure and meteorology instrument, and colleagues in their published report.

High in the atmosphere, at altitudes of 80 kilometers (50 miles) above the surface, temperatures were cold enough to make carbon dioxide condense and form carbon dioxide clouds. At altitudes of between 60 kilometers and 120 kilometers (37 miles and 75 miles), the Martian atmosphere was an average of 20 degrees colder than Viking measurements, Schofield reports. Seasonal variations and Pathfinder's entry at 3 a.m. local solar time, compared with Viking's entry at 4 p.m. local solar time, may account for these variations. On the surface, however, daytime temperatures were typically 10 to 12 degrees warmer than Viking surface temperatures.

Mars Pathfinder measured regular pressure fluctuations twice a day, which suggested that a moderate amount of dust is being uniformly mixed in a warm lower atmosphere, as was the case with Viking data. The daily average pressure reached a minimum on the 20th day of the mission (Sol 20), indicating the winter south polar cap had reached its maximum size.

Schofield reports that surface temperatures follow a regular daily cycle, with a maximum of 15 degrees Fahrenheit during the day and a minimum of minus 105 degrees Fahrenheit at night. The science team also observed rapid daytime temperature fluctuations of up to 30 degrees Fahrenheit in as little as 25 to 30 seconds. These observations suggest that cold air was warmed by the surface and convected upward in small eddies.

Pathfinder encountered winds that were light and variable compared to the Viking landers, Schofield reports. The winds blew steadily from the south during the Martian nights, but during the day they rotated in a clockwise direction from south to west to north to east. Whirlwinds or dust devils were detected repeatedly from midmorning through the late afternoons.

Other scientific findings of the Mars Pathfinder mission are:

- Chemical analyses returned by Mars Pathfinder indicate some rocks appear to be high in silica, suggesting differentiated parent materials. These rocks are distinct from the meteorites found on Earth that are thought to be of Martian origin.

- The identification of rounded pebbles and cobbles on the ground, and sockets and pebbles in some rocks, suggests conglomerates that formed in running water, during a warmer past in which liquid water was stable.

- The measurement of the moment of inertia of Mars by tracking Pathfinder radio data indicates the radius of the central metallic core is greater than 1300 km but less than roughly 2000 km.

- Airborne dust is magnetic with a mean size of about 1 micron. Interpretations suggest the magnetic mineral is maghemite, which may have been freeze- dried on the particles as a stain or cement, and that the iron may have been leached out of crustal materials by an active hydrologic cycle.

- Remote-sensing data at a scale of generally greater than 1 kilometer and an Earth analog correctly predicted a rocky plain safe for landing and roving, with a variety of rocks deposited by catastrophic floods that are relatively dust free.

- Imaging revealed early morning water ice clouds in the lower atmosphere, which sublimate away as the atmosphere warms.

- Abrupt temperature fluctuations with time and height were recorded in the morning, which was consistent with warming of the atmosphere by the surface and convected upwards in small eddies into the atmosphere.

- Dust devils were frequently measured by temperature, wind and pressure sensors, and at least one likely contained dust, suggesting that these gusts are a mechanism for mixing dust into the atmosphere.

- The soil chemistry of Ares Vallis appears to be similar to that of the Viking 1 and 2 landing sites, suggesting that the soil may be a globally deposited unit.

- Some rocks at the landing site appear grooved and fluted, suggesting abrasion by saltating sand-sized particles. Dune-shaped deposits were also found in a trough behind the Rock Garden, indicating the presence of sand.

- The weather was similar to the weather encountered by Viking 1; there were rapid pressure and temperature variations, downslope winds at night and light winds in general. Temperatures at the surface were about 10 degrees Kelvin warmer than those measured by Viking 1.

- The atmosphere has been a pale pink color due to fine dust mixed in the lower atmosphere, as was seen by Viking. Particle size and shape estimates and the amount of water vapor in the atmosphere are also similar to that measured by Viking.

Courtesy JPL/NASA

Appendix H

Pictures Boost View That Mars Supported Life

NASA Orbiter spots evidence that liquid water lasted eons

Robert S. Boyd

WASHINGTON — The odds that Mars once supported life have been strengthened by a set of spectacular new pictures from a NASA spaceship circling our neighboring planet.

The latest images indicate that liquid water probably existed on Mars for at least a million years—long enough for primitive organisms to form—according to Wesley Huntress Jr., NASA's associate administrator for space science.

Earlier evidence sent back by the Mars *Pathfinder* lander last year confirmed that an enormous flood took place on Mars a billion or more years ago, but did not reveal how long the water remained on the now bone-dry surface.

"This is the first evidence of long-term water flow, not just a catastrophic event." Huntress told a meeting in Philadelphia of the American Association for the Advancement of Science.

In recent years, scientists have come to think that life is possible—perhaps probable—wherever there is liquid water, even in conditions of extreme heat, cold or darkness.

Understanding how life might have gotten an early toehold on Mars—and why it apparently disappeared—could help explain the origin of life on Earth and the dangers of environmental change.

The new pictures were returned by the *Mars Global Surveyor*, a science satellite that began orbiting the red planet last September. It will spend the next two years taking detailed pictures of the entire Martian surface in preparation for future landings.

One *surveyor* image shows what appears to be a dried riverbed inside a 1 1/2 mile wide canyon known as Nanedi Vallis, part of a series of valleys cutting through plains in the Xanthe Terra region near the Martian equator, about 1,000 miles southwest of the *Pathfinder* landing site.

The channel meanders back and forth over the canyon floor, bending, looping and following different paths, much like the Mississippi River.

"This is clear evidence of a fast-flowing (river) that changed its course several times," Huntress said. "It probably took a million or so years to form."

Another picture shows terraces along the base of

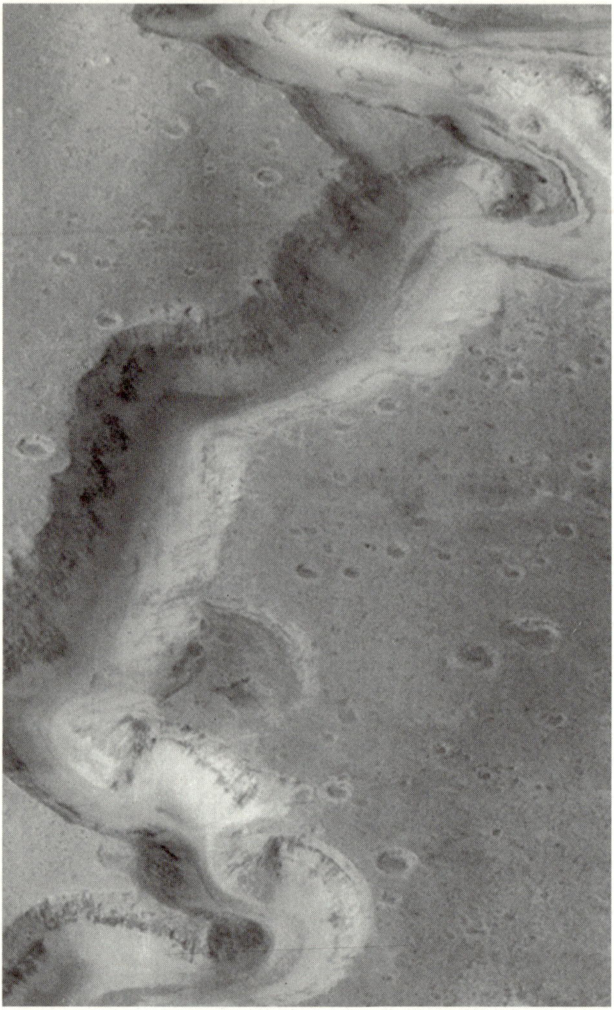

"NASA/JPL/Caltech"

This picture returned by the *Mars Global Surveyer* shows a dried riverbed inside a 1 1/2 mile wide canyon, Nanedi Vallis, part of a series of valleys cutting through plains in the Xander Terra region near the Martian equator, about 1,000 miles southwest of the *Pathfinder* landing site.

mighty cliffs lining another valley, 600 miles farther south, known as Vallis Marineris. Huntress said they resemble the terraces formed by the Colorado River as it carved the Grand Canyon in Arizona.

Other images reveal what appear to be dried-up lake beds, like the landscape in the American Southwest, and a multiplicity of fine-grained sand dunes, further evidence of long-vanished water.

The *Global Surveyor* also sent back fresh pictures of the Martian south pole, where NASA's *Polar Lander* will touch down in December 1999. They show bands of eroded layered rock, reminiscent of the edges of Alaskan ice sheets.

The *Polar Lander* will spend three months searching for traces of subsurface water in the frozen terrain. Its package of scientific instruments include two "microprobes" designed to dig three feet below the surface and report what they find.

Recent discoveries on Earth—under the ocean, at the bottom of oil wells, in boiling hot springs and acid pools—show that simple life forms can thrive in extreme environments.

"We used to think that life was fragile," Huntress said. "But wherever liquid water and chemical energy are found, there is life. There is no exception. Life may be a cosmic imperative."

According to NASA officials, whether life once existed on Mars—or still does—probably cannot be answered until a spaceship returns samples of rocks and soil to Earth for detailed study. Such a mission is planned to depart Earth in 2005 and return in 2008.

Scientists are already studying ways to keep Martian organisms isolated so they won't spread and infect the Earth.

Doubts are growing that the famed meteorite, ALH84001, picked up in the Antarctic actually contains fossils of Martian organisms, as reported in 1996.

Some scientists say the fossils are really inorganic artifacts created during the meteorite's fiery ride through space. Others suggest the potato-sized rock was contaminated by Earth organisms.

"It's very difficult to come to a consensus," said Klaus Kiel, a physicist at the Hawaiian Institute of Geophysics and Planetology.

Appendix I

New Global Surveyor Data

reveals deeply layered terrain, magnetic features, and genesis of a Martian dust storm

Diane Ainsworth

For the first time in Mars exploration, a spacecraft has captured the full evolution of a Martian dust storm. NASA's Mars Global Surveyor mission also has returned new insights into the deeply layered terrain and mineral composition of the Martian surface, and to highly magnetized crustal features that provide important clues about the planet's interior.

This first set of formal results comes from data obtained in October and November 1997, while the spacecraft was just beginning to use the drag of Mars' upper atmosphere to lower and circularize its orbit in a process called aerobraking. At the time, a dust storm was brewing on Mars and had grown to about the size of the South Atlantic Ocean.

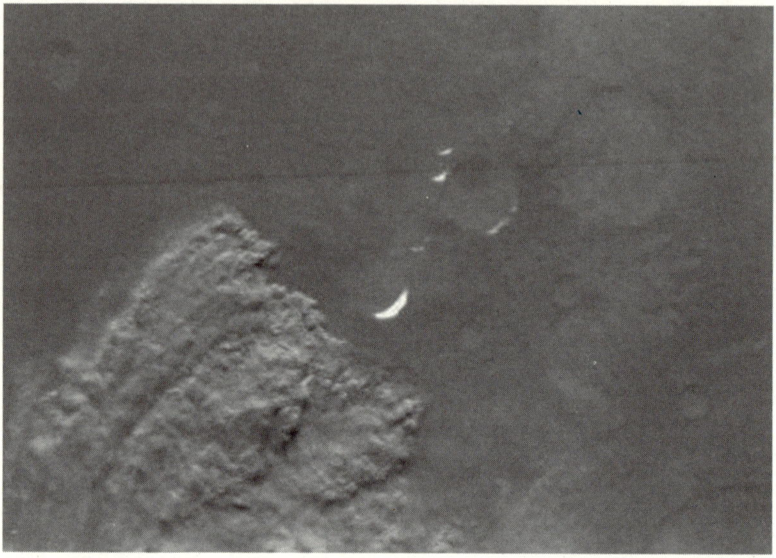

"NASA/JPL/Caltech"

Dust storms are relatively common on Mars. They tend to occur in high topographic and/or high thermal gradients (usually near the caps), where surface winds would be strongest. This storm is hundreds of kilometers in extent and is located near the edge the south polar cap. Los Alamos Nat. Lab. Processing by Calvin Hamilton.

The Global Surveyor data suggest that it began as a set of small dust storms along the edge of the planet's southern polar cap, according to Dr. Arden Albee of the California Institute of Technology, Pasadena, California, the Mars Global Surveyor mission scientist. By Thanksgiving, it had expanded into a large regional dust storm in Noachis Terra that covered almost 180 degrees longitude, while spanning 20 degrees south latitude to nearly the tip of the Martian equator.

"As this storm obscured the Martian landscape, we followed it in detail using several instruments onboard Mars Global Surveyor," Albee said. "The thermal emission spectrometer mapped the temperature and opacity of the atmosphere while the camera followed the visual effects. The effects of the storm extended to great heights of about 130 kilometers (80 miles) and resulted in great increases in both atmospheric density and variability from orbit to orbit. These atmospheric measurements have great significance for future Mars missions that will be using aerobraking techniques too."

Before the storm, atmospheric dust was generally distributed very uniformly, Albee said. Observations of the limb of the planet in the northern hemisphere revealed both low-lying dust hazes and detached water-ice clouds at altitudes of up to 55 kilometers (34 miles). Movement of these clouds was tracked by the spectrometer as the planet rotated. Atmospheric turbulence disrupted these cloud patterns as the small storms began to rise and kick more dust into the air. As the storm began to abate, small local storms began to crop up again along the edges of the south polar cap, and ice clouds formed in depressions as the carbon dioxide cap continued to retreat.

In addition to these unprecedented observations of a full-blown Martian dust storm, measurements from the spacecraft's magnetometer and electron reflectometer have yielded new findings about Mars' strong, localized magnetic fields. These patches of the crust, which register high levels of magnetism, are beginning to unlock some of the mysteries surrounding Mars' internal dynamo and when it died, said Dr. Mario Acuna of NASA's Goddard Space Flight Center, Greenbelt, Maryland.

"These locally magnetized areas on Mars could not form without the presence of an overall global magnetic field that was perhaps as strong as Earth's is today," says Acuna. "Since the internal dynamo that powered the global field is extinct, these local magnetic fields act as fossils, preserving a record of the geologic history and thermal evolution of Mars."

Magnetic fields are created by the movement of electrically conducting fluids, and a planet can generate a global magnetic field if its interior consists of molten metal hot enough to undergo convective motion, similar to the churning motion seen in boiling water.

"The small size and highly magnetic nature of these crustal features, which measure on the order of 50 kilometers (30 miles), are found within the ancient cratered terrain rather than within the younger volcanic terrain," Acuna said. "By correlating crustal age with magnetization, we have a perfect window on Mars' past, which will help us to determine when Mars' internal dynamo ceased operating."

High-resolution images of dunes, sandsheets and drifts also are helping reveal earlier chapters of Martian history. Landforms shaped by erosion are almost everywhere, according to Albee, and many bear a striking resemblance to Colorado's Rocky Mountains. Rocky ridges poke through the Martian dust just as the jagged edges of

Mars update, May 27, 1998

Global Surveyor Data Reveals More Evidence of Abundant Water, Thermal Activity in Mars' Past

Diane Ainsworth

New mineralogical and topographic evidence suggesting that Mars had abundant water and thermal activity in its early history is emerging from data gleaned by NASA's Mars Global Surveyor spacecraft.

Scientists are getting more glimpses of this warmer, wetter past on Mars while Global Surveyor circles the planet in a temporary 11.6-hour elliptical orbit. Findings from data gathered during the early portions of this hiatus in the mission's orbital aerobraking campaign are being presented today at the spring meeting of the American Geophysical Union in Boston.

Among many results, the Thermal Emission Spectrometer instrument team, led by Dr. Philip Christensen of Arizona State University, Tempe, has discovered the first clear evidence of an ancient hydrothermal system. This finding implies that water was stable at or near the surface and that a thicker atmosphere existed in Mars' early history.

Measurements from the spectrometer show a remarkable accumulation of the mineral hematite, well-crystallized grains of ferric (iron) oxide that typically originate from thermal activity and standing bodies of water. This deposit is localized near the Martian equator, in an area approximately 300 miles (500 kilometers) in diameter.

Fine-grained hematite, with tiny particles no larger than specks of dust, generally forms by the weathering of iron-bearing minerals during oxidation, or rusting, which can occur in an atmosphere at low temperatures. The material has been previously detected on Mars in more dispersed concentrations and is widely thought to be an important component of the materials that give Mars its red color. The presence of a singular deposit of hematite on Mars is intriguing, however, because it typically forms by crystal growth from hot, iron-rich fluids. Meanwhile, the Mars Orbiter Laser Altimeter instrument is giving mission scientists their first three-dimensional views of the planet's north polar ice cap. Principal Investigator Dr. David Smith of NASA's Goddard Space Flight Center, Greenbelt, MD, and his team have been using the laser altimeter to obtain more than 50,000 measurements of the topography of the polar cap in order to calculate its thickness, and learn more about related seasonal and climatic changes.

These initial profiles have revealed an often striking surface topology of canyons and spiral troughs in the water and carbon dioxide ice that can reach depths as great as 3,600 feet below the surface. Many of the larger and deeper troughs display a staircase structure, which may ultimately be correlated with seasonal layering of ice and dust observed by NASA's Viking mission orbiters in the late 1970s.

The laser data also have shown that large areas of the ice cap are extremely smooth, with elevations that vary only a few feet over many miles. At 86.3 degrees north, the highest latitude yet sampled, the cap achieves an elevation of 6,600 to 7,900 feet (1.25 to 1.5 miles or 2-2.5 kilometers) over the surrounding terrain. The laser measurements are accurate to approximately one foot (30 centimeters) in the vertical dimension.

In June, the ice cap's thickness will reach a maximum during the peak of the northern winter season. Thickness measurements from April will be compared to those that will be taken in June, contributing to a greater understanding of the Martian polar cap's formation and evolution.

In addition, the Global Surveyor accelerometer team, led by Dr. Gerald Keating of George Washington University, Washington, DC, has discovered two emormous bulges in the upper atmosphere of Mars in the Northern hemisphere, on opposite sides of the planet near 90 degrees east latitude and 90 degrees west longitude. These bulges rotate with the planet, causing variations of nearly a fractor of two in atmospheric pressure, and systematic variations in altitude of a given constant pressure of about 12,000 feet (four kilometers).

After a month-long period during which the Sun was between Earth and Mars and thus degraded communications with Global Surveyor, the spacecraft has resumed taking scientific data in its temporary elliptical orbit. In September, it will once again begin dipping into the upper atmosphere of Mars each orbit in a process called aerobraking. The drag from this procedure will allow the spacecraft to reach a low circular orbit and begin its primary two-year global mapping mission starting in March 1999.

Courtesy JPL/NASA

"This is the first evidence of long-term water flow, not just a catastrophic event." Huntress told a meeting in Philadelphia of the American Association for the Advancement of Science.

In recent years, scientists have come to think that life is possible—perhaps probable—wherever there is liquid water, even in conditions of extreme heat, cold or darkness.

Understanding how life might have gotten an early toehold on Mars—and why it apparently disappeared—could help explain the origin of life on Earth and the dangers of environmental change.

The new pictures were returned by the *Mars Global Surveyor*, a science satellite that began orbiting the red planet last September. It will spend the next two years taking detailed pictures of the entire Martian surface in preparation for future landings.

One *surveyor* image shows what appears to be a dried riverbed inside a 1 1/2 mile wide canyon known as Nanedi Vallis, part of a series of valleys cutting through plains in the Xanthe Terra region near the Martian equator, about 1,000 miles southwest of the *Pathfinder* landing site.

The channel meanders back and forth over the canyon floor, bending, looping and following different paths, much like the Mississippi River.

"This is clear evidence of a fast-flowing (river) that changed its course several times," Huntress said. "It probably took a million or so years to form."

Another picture shows terraces along the base of

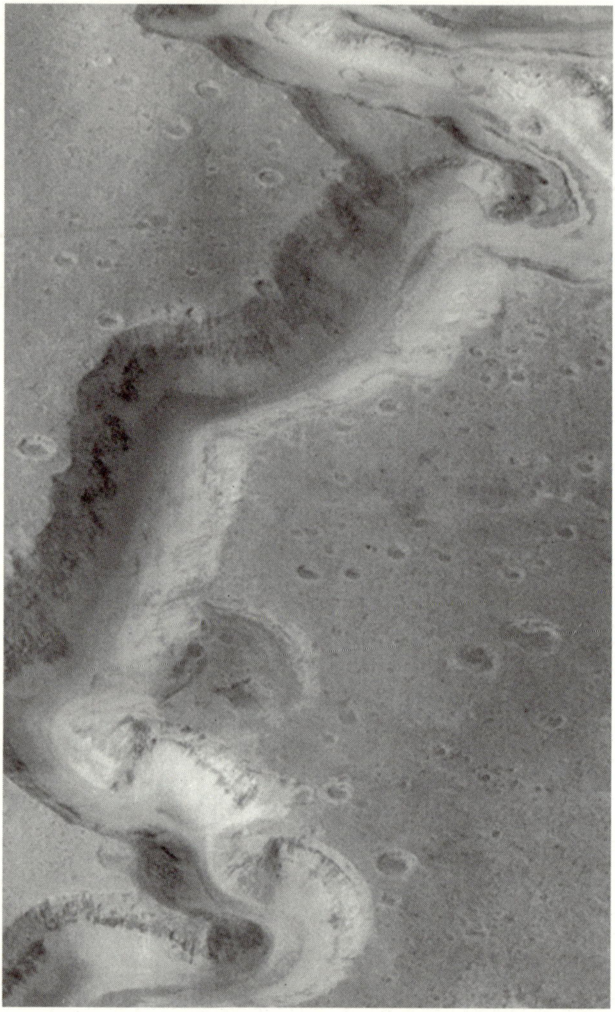

"NASA/JPL/Caltech"

This picture returned by the *Mars Global Surveyer* shows a dried riverbed inside a 1 1/2 mile wide canyon, Nanedi Vallis, part of a series of valleys cutting through plains in the Xander Terra region near the Martian equator, about 1,000 miles southwest of the *Pathfinder* landing site.

mighty cliffs lining another valley, 600 miles farther south, known as Vallis Marineris. Huntress said they resemble the terraces formed by the Colorado River as it carved the Grand Canyon in Arizona.

Other images reveal what appear to be dried-up lake beds, like the landscape in the American Southwest, and a multiplicity of fine-grained sand dunes, further evidence of long-vanished water.

The *Global Surveyor* also sent back fresh pictures of the Martian south pole, where NASA's *Polar Lander* will touch down in December 1999. They show bands of eroded layered rock, reminiscent of the edges of Alaskan ice sheets.

The *Polar Lander* will spend three months searching for traces of subsurface water in the frozen terrain. Its package of scientific instruments include two "microprobes" designed to dig three feet below the surface and report what they find.

Recent discoveries on Earth—under the ocean, at the bottom of oil wells, in boiling hot springs and acid pools—show that simple life forms can thrive in extreme environments.

"We used to think that life was fragile," Huntress said. "But wherever liquid water and chemical energy are found, there is life. There is no exception. Life may be a cosmic imperative."

According to NASA officials, whether life once existed on Mars—or still does—probably cannot be answered until a spaceship returns samples of rocks and soil to Earth for detailed study. Such a mission is planned to depart Earth in 2005 and return in 2008.

Scientists are already studying ways to keep Martian organisms isolated so they won't spread and infect the Earth.

Doubts are growing that the famed meteorite, ALH84001, picked up in the Antarctic actually contains fossils of Martian organisms, as reported in 1996.

Some scientists say the fossils are really inorganic artifacts created during the meteorite's fiery ride through space. Others suggest the potato-sized rock was contaminated by Earth organisms.

"It's very difficult to come to a consensus," said Klaus Kiel, a physicist at the Hawaiian Institute of Geophysics and Planetology.

Appendix I

New Global Surveyor Data

reveals deeply layered terrain, magnetic features, and genesis of a Martian dust storm

Diane Ainsworth

For the first time in Mars exploration, a spacecraft has captured the full evolution of a Martian dust storm. NASA's Mars Global Surveyor mission also has returned new insights into the deeply layered terrain and mineral composition of the Martian surface, and to highly magnetized crustal features that provide important clues about the planet's interior.

This first set of formal results comes from data obtained in October and November 1997, while the spacecraft was just beginning to use the drag of Mars' upper atmosphere to lower and circularize its orbit in a process called aerobraking. At the time, a dust storm was brewing on Mars and had grown to about the size of the South Atlantic Ocean.

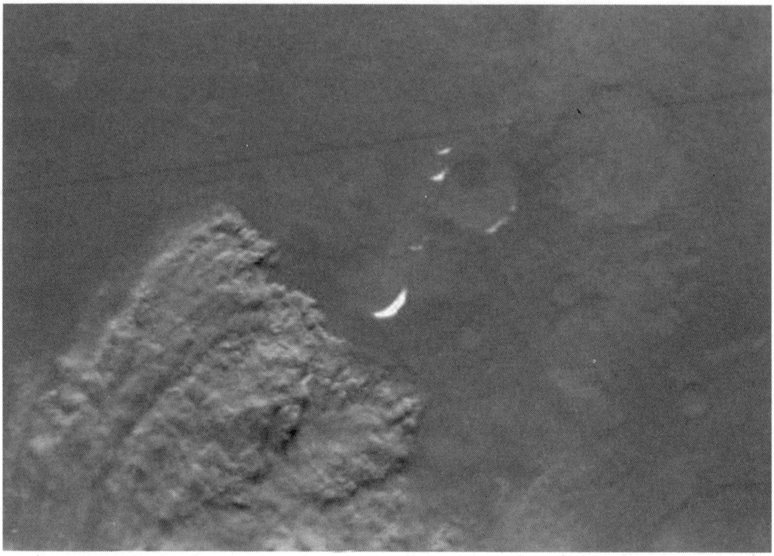

"NASA/JPL/Caltech"

Dust storms are relatively common on Mars. They tend to occur in high topographic and/or high thermal gradients (usually near the caps), where surface winds would be strongest. This storm is hundreds of kilometers in extent and is located near the edge the south polar cap. Los Alamos Nat. Lab. Processing by Calvin Hamilton.

The Global Surveyor data suggest that it began as a set of small dust storms along the edge of the planet's southern polar cap, according to Dr. Arden Albee of the California Institute of Technology, Pasadena, California, the Mars Global Surveyor mission scientist. By Thanksgiving, it had expanded into a large regional dust storm in Noachis Terra that covered almost 180 degrees longitude, while spanning 20 degrees south latitude to nearly the tip of the Martian equator.

"As this storm obscured the Martian landscape, we followed it in detail using several instruments onboard Mars Global Surveyor," Albee said. "The thermal emission spectrometer mapped the temperature and opacity of the atmosphere while the camera followed the visual effects. The effects of the storm extended to great heights of about 130 kilometers (80 miles) and resulted in great increases in both atmospheric density and variability from orbit to orbit. These atmospheric measurements have great significance for future Mars missions that will be using aerobraking techniques too."

Before the storm, atmospheric dust was generally distributed very uniformly, Albee said. Observations of the limb of the planet in the northern hemisphere revealed both low-lying dust hazes and detached water-ice clouds at altitudes of up to 55 kilometers (34 miles). Movement of these clouds was tracked by the spectrometer as the planet rotated. Atmospheric turbulence disrupted these cloud patterns as the small storms began to rise and kick more dust into the air. As the storm began to abate, small local storms began to crop up again along the edges of the south polar cap, and ice clouds formed in depressions as the carbon dioxide cap continued to retreat.

In addition to these unprecedented observations of a full-blown Martian dust storm, measurements from the spacecraft's magnetometer and electron reflectometer have yielded new findings about Mars' strong, localized magnetic fields. These patches of the crust, which register high levels of magnetism, are beginning to unlock some of the mysteries surrounding Mars' internal dynamo and when it died, said Dr. Mario Acuna of NASA's Goddard Space Flight Center, Greenbelt, Maryland.

"These locally magnetized areas on Mars could not form without the presence of an overall global magnetic field that was perhaps as strong as Earth's is today," says Acuna. "Since the internal dynamo that powered the global field is extinct, these local magnetic fields act as fossils, preserving a record of the geologic history and thermal evolution of Mars."

Magnetic fields are created by the movement of electrically conducting fluids, and a planet can generate a global magnetic field if its interior consists of molten metal hot enough to undergo convective motion, similar to the churning motion seen in boiling water.

"The small size and highly magnetic nature of these crustal features, which measure on the order of 50 kilometers (30 miles), are found within the ancient cratered terrain rather than within the younger volcanic terrain," Acuna said. "By correlating crustal age with magnetization, we have a perfect window on Mars' past, which will help us to determine when Mars' internal dynamo ceased operating."

High-resolution images of dunes, sandsheets and drifts also are helping reveal earlier chapters of Martian history. Landforms shaped by erosion are almost everywhere, according to Albee, and many bear a striking resemblance to Colorado's Rocky Mountains. Rocky ridges poke through the Martian dust just as the jagged edges of cliffs pierce through a blanket of snow in the Rockies. Martian dust appears to have spilled down the sides of ridges just as fresh snow slides down a ski slope.

"One almost expects to see ski tracks crisscrossing the area," Albee added. "These images present a sharp contrast to the images of boulder-strewn deserts found at the Viking and Pathfinder landing sites."

The Martian crust also exhibits much more layering at great depth than was expected. The steep walls of canyons, valleys and craters show the Martian crust to be stratified at scales of a few tens of yards, which is an exciting discovery, Albee noted. "At this point we simply do not know whether these layers represent piles of volcanic flows or sedimentary rocks that might have formed in a standing body of water," he said.

The thermal emission spectrometer, led by principal investigator Dr. Philip Christensen of Arizona State University, is beginning to obtain a few infrared emission spectra of the surface, although it is still too cold on the surface for the best results. The best spectra clearly indicate the presence of pyroxene and plagioclase, minerals which are common in volcanic rocks, with a variable amount of dust component. No evidence was found for carbonate minerals, clay minerals or quartz. If present in these rocks, their abundance must be less than about 10 percent.

Their absence indicates that carbonates are not ubiquitous over the surface of the planet, but they may still be found in specific locations that either favored their initial deposition or their subsequent preservation. This finding could have important implications for identifying areas that may preserve signs of ancient life on Mars, since carbonate minerals are commonly formed in biological processes, Albee said.

Striking results also have been obtained from Global Surveyor's laser altimeter over Mars' northern hemisphere, which is exceptionally flat with slopes and surface roughness increasing toward the equator, according to principal investigator Dr. David Smith of NASA Goddard Space Flight Center, Greenbelt, Maryland. The initial data for this region helps scientists interpret a variety of landforms, including the northern polar cap, gigantic canyons, ridges, craters of all sizes and shield volcanoes. Most surprising are views of extraordinarily mundane regions—as flat as the Bonneville Salt Flats in Utah—that extend over vast northern regions of the planet.

Mars Global Surveyor will complete the first phase of its two-part aerobraking strategy at the end of March, at which time the science instruments will be turned on again for most of the next six months. Over this period, the spacecraft will stay in an 11 1/2-hour orbit and collect an additional bounty of data at a closest approach of about 170 kilometers (106 miles) above the surface, much closer than the spacecraft will pass over the planet once it has reached its formal mapping orbit in March 1999. This closer orbit will allow the science teams to take more detailed measurements of the Martian atmosphere and surface without magnetic interference from the solar wind.

"When we decided to slow the pace of aerobraking

to reduce the force on the solar panel that was damaged after launch, we knew we would get a bonus—the ability to collect much more science data closer to the planet than will be possible during the prime mapping mission," said Glenn E. Cunningham, Mars Global Surveyor project manager at NASA' Jet Propulsion Laboratory, Pasadena, California. "Additionally, the six-month period between the end of March and early September will yield an extraordinary opportunity as the lowest point of the orbit migrates over the northern polar cap. All of this information that is coming back now is really icing on the cake, a spectacular precursor to the global mapping data expected to start flowing next year."

Pathfinder & Viking landing sites & the Cydonia region

NASA's Mars Global Surveyor spacecraft is about to begin a summer-long set of scientific observations of the red planet from an interim elliptical orbit, including several attempts to take images of features of public interest ranging from the Mars Pathfinder and Viking mission landing sites to the Cydonia region.

In addition, the Mars Pathfinder and Viking landers are very small targets to image, even at the closest distance possible, because they are the smallest objects that the camera can see. The Cydonia features, on the other hand, are hundreds to thousands of times larger and the camera should be able to capture some of the features in that area.

Global Surveyor's observations of the Viking and

Pathfinder landing sites will provide scientists with important information from which to tie together surface observations and orbital measurements of the planet. Data from landing sites provide "ground truth" for observations of the planet made from space.

As for the "Face on Mars" feature, "Most scientists believe that everything we've seen on Mars is of natural origin," said Dr. Carl Pilcher, acting science director for solar system exploration in NASA's Office of Space Science, Washington, DC. "However, we also believe it is appropriate to seek to resolve speculation about features in the Cydonia region by obtaining images when it is possible to do so."

Courtesy JPL/NASA

CONCLUSION

Conclusion

Definite conclusions about the possibility of Martian life will have to wait until future spacecraft bring carefully selected samples back from the planet. The first of ten flights of orbiting spacecraft and robotic landers to explore Mars began in December 1996 with the launching of *Mars Pathfinder* and the release of its micro rover, *Sojourner,* on July 5, 1997.

It is our purpose in presenting this pioneering information about the reality of extraterrestrial life to take the discussion of extraterrestrial life out of the limbo of science fiction.

As the twentieth century comes to a close, we will begin to see, in effect, a wider window into the universe and, more specifically, an opening of the "cosmic curtain" that heretofore hid the people of planet Earth from the reality of other terrestrial civilizations in our galaxy.

Much new information is forthcoming, not only from

the material published by the Unarius Academy of Science, but from individuals who have had personal experience in their contact with other human beings living on outlying planets throughout our universe. It is hoped, therefore, that the reader will be reawakened to the richness of the cosmic design of life and gain a greater understanding of the intelligence that pervades life, both on the microscopic and on the macroscopic scale. Most importantly, our prehistory has provided us with trace elements of visitors from other planets and solar systems, and indeed from constellations in our own backyard, the Milky Way Galaxy.

We are just beginning to see the incredible substance of our cosmic environment and the potential that beckons to each individual. It is a potential that is rich in the stuff of the mind and speaks the truth about one's spiritual purpose in the cyclic, evolutionary passage throughout life, life after life.

The inescapable logic and reason for previous probes to planet Mars in 1964 and 1976, the *Mariner* and *Viking* probes, was to give to the scientific community a richer picture of our neighboring planet. However, this information has only begun to satisfy our interest in the possibility of biological life on Mars and in what may have been an ancient ecosystem on the fourth planet from the sun. It is inescapable that there is a history that touches the core of many people who, beginning with Percival Lowell, had a mission to verify the actuality of human origins on planet Mars. When additional information from not only the *Pathfinder* but future probes launched by NASA is examined, we will realize that planet Earth is not the sole proprietor of Homo sapiens.

Throughout our history, and from records of the civilization of Atlantis, there have been records of scientific research, left by individuals in different disciplines of knowledge, which have formed a tapestry depicting the prehistory of humankind on Earth. This history has been woven into the mythos traceable to varied cultures and societies throughout our planet.

The eyewitness account of Ernest L. Norman is invaluable because the author verifies the power of the mind as an interdimensional communication device. The fact is we humans have proven that we have the ability to invent computers and television devices, which are derived from a higher intelligence. Therefore, it is not too difficult to realize that the deep and abiding interest we have in researching the possibility of life on planet Mars and other extraterrestrial solar planets is a product of our past. In other words, memory contains the substance of previous life experiences.

The focus, therefore, of this book is the developmental syllabus of the mind, which serves as man's research vehicle to explore the microcosm of himself and the macrocosm of the universe in which he lives.

History reveals that mankind has always been searching for the understanding of life, indeed, yearning for knowledge of the cosmic scheme and our position in the design called Life. Planet Earth is, in reality, a research laboratory for the billions of people who inhabit its surface and the hundreds of millions of other specimens—plants, animals, insects, birds, and the watery ecosystem that provides the environmental basis for survival on Earth.

We humans, curious about the unknown which lies beyond us, have looked outwardly from our small planet, longing for knowledge of other planets in the immensity of space. It may be because of our inquiring mind that we are always seeking for the logic and reason for our existence in an immense universe and for our purpose in the Cosmos!

It is not too much to conjecture, therefore, that in the very active inquiry into the substance of life, no matter on what level, microcosm or macrocosm, the evidence undeniably suggests that there is more to the hard copy of physical life.

The subject of this book *The Truth About Mars* illustrates and expands on the truth about extraterrestrial life. How did it come about? The desire for a logic and reason for the existence of life validates that the human mind is, itself, constructed from the stuff of the universe, the energy that has fashioned the planets, stars, galaxies, universes and humankind.

The cosmic curtain is indeed opening as we come to the conclusion of the two-thousand-year millennium, which itself is a subcycle of the sixth cycle of the recessional. After all, life is for learning, and learning is indeed one's life-long purpose, to advance the logic and reason for one's involvement in life and to further the purpose of progressive evolution.

About the Author

Ernest L. Norman made his debut into this planet Earth in 1904, in a small town in northern Utah. It was apparent right from the first that he was an unusual child. His mother nearly died in the process of his introduction because of his abnormally large head. It is said he had the body of an eight-pound child but weighed over twelve pounds!

Before he was barely two, he was experimenting with writing, and long before he went to school for the first time, he was quite familiar with the English language, so much so, that he was reading his father's library. His father, incidentally, was a very learned man of royal Norwegian descent, who had degrees in law, psychology, physiology, and phrenology.

The author was the fifth of eight boys and girls, all strong lads and lassies, and it was quite natural for them to resent having a brother who was so studious.

At the tender age of five he constructed his first microscope using the eyepiece section from his father's telescope. By inserting it in a wooden frame made from a cigar box and a small piece of mirror, he was able to count the hairs on earthworms. At the age of six he performed an unusual and prodigious feat. Using his knowledge of Archimedean laws of fulcrums, levers and rollers, he moved an eight-by-twelve-foot coal shed, containing one-half a ton of coal, over a distance of approximately 200 feet, through an apple orchard and over soft ground to a new and more convenient location. This feat took him about three weeks to accomplish, and it was one that would have taxed the strength and endurance of a strong man. This accomplishment was carefully noted, day by day, by his father who would boastingly report the progress made to the townspeople.

It was also during this early period of life that he constructed a rabbit hutch which was vastly superior in design and workmanship to one constructed by an adult neighbor more than six times his age. This he did, using old rusted out, discarded tools.

Another time, at the age of seven, he bested his father in an argument i.e., that all energy was electronic. At present he is completely vindicated inasmuch as science today is resolving into this conclusion.

During his early teenage years in junior and senior

high school, he established several new "high water" marks in biology, genetics, science, etc., and won several noteworthy citations as well as attracting some interest from his teachers.

It is estimated that at the age of fourteen he had a vocabulary of about sixty thousand words! It was easy to see then, that this boy, who in wintertime read almost continuously or dreamed the summer away watching nature, had not wasted or played his time away as most lads are wont to do.

At the age of seventeen his family moved to California and temporarily, at least, his formal schooling was finished. But he persisted, even taking night classes in various subjects. At the age of twenty-three, and just before the depression, he married and remained so for fourteen years.

During this time he became very active in radio and electronics. It was his wife who always said they had the best radios in the neighborhood, and they were always hand constructed.

After World War II, he devoted himself to his lifetime dream, metaphysics. From the very start, in this work, it became evident he possessed an outstanding clairvoyant development. During the war years, he demonstrated this talent not only in churches and lecture halls but in almost any place opportunity presented itself, and he achieved no small measure of fame in this capacity.

However, it was his ambition at this time, as it has been for many years, to fill in the gaps in our scientific and spiritual philosophies of the world and to set up a new and integrated philosophy of life.

With this most outstanding ability of clairvoyance, coupled with a tremendous grasp of scientific knowledge, he was very humble and unpretentious, refusing to attach the stigma of self to whatever came through the channel of his mind. He was ever aware of his attunement with the Superconscious.

Selected Bibliography

Antoniadi, E. M. *La planete Mars*, 1659-1929. Paris: Hermann et Cie, 1930, English translation: *The Planet Mars.* Trans. Patrick Moore. Shaldon, Devon, England: Keith Reid, 1975.

Baker, V. R. *The Channels of Mars.* Austin, Texas: University of Texas Press, 1982.

Bradbury, Ray, Arthur C. Clarke, Bruce Murray, Carl Sagan, and Walter Sulliven. *Mars and the Mind of Man.* New York: Harper and Row, 1973.

Carr, M. H. *The Surface of Mars*, New Haven, Conn.: Yale University Press, 1981.

Cotardiere, Philippe de la, and Patrick Fuentes. *Camille Flammarion.* Paris: Flammarion, 1994.

Davis, Paul. *Are We Alone?* New York: John Wiley & Sons. 1988

Digregorio, Barry E., Gilbrt V. Levin, & Patricia Ann Stratt. *Mars: The Living Planet.* Berkeley, California: Frog Ltd., 1997

Drake, Frank and Dava Sobel. *Is Anyone out There?* The Scientific Search for Extraterrestrial Intelligence. N. Y.: Delacorte Press, 1993

Flammarion, Camille. *La planete Mars et ses conditions d'habitailite.* 2 vols. Paris: Gauthier-Villars et Fils, 1892 and 1909.

Goldsmith, Donald. *The Hunt for Life on Mars.* New York: Dutton Books, 1997.

Hartmann, W. K., and Odell Raper. *The New Mars: The Discoveries of Mariner 9.* NASA SP-337. Washington, D.C.: National Aeronautics and Space Administration, 1974.

Hoyt, W.G. *Lowell and Mars.* Tucson, Arizona: University of Arizona Press, 1976.

Kieffer, H. H., B. M. Jakosky, C. W. Snyder, and M. S. Matthews. *Mars.* Tucson: University of Arizona Press, 1992.

Lowell, Percival. *Mars.* Boston: Houghton Mifflin, 1895.

___*Mars and Its Canals.* New York: Macmillan, 1906.

___*Mars as the Abode of Life.* New York: Macmillan, 1910.

Mutch, T.A., R. E. Arvidson, J. W. Head, K. L. Jones, and R.S. Saunders. *The Geology of Mars.* Princeton, NJ: Princeton University Press, 1976.

Norman, Ruth E., *Mars, Underground Cities Discovered.* El Cajon, California: Unarius Publications, 1977.

___*Interdimensional Physics: The Mind and the Universe.* El Cajon, California: Unarius Publications, 1989

Norman, Ruth E. & Charles Spiegel, *Communications From Outer Space.* El Cajon, California: Unarius Publications, 1984.

___*Preparation for the Landing.*El Cajon, California: Unarius Publications, 1987.

Richardson, R.S., and Chesley Bonestell. *Mars.* New York: Harcourt, Brace and World, 1964.

Slipher, E. C. *A Photographic History of Mars.* Flagstaff, Arizona: Northland Press, 1962.

Vaucouleurs, G. de. *The Planet Mars.* Trans. P. Moore. London: Faber and Gaber, 1950.

Physics of the Planet Mars: An Introduction to Areophysics. Trans. P. Moore. London: Faber and Faber, 1954.

Viking Lander Imaging Team. *The Martian Landscape.* NASA SP-425. Washington, D.C.: NASA, 1978.

Wallace, Alfred Russell. *Is Mars Inhabited?* N.Y: Macmillan, 1907.